VISUAL PATTERNS

IN

PASCAL'S

TRIANGLE

Dale Seymour

DALE SEYMOUR PUBLICATIONS

Illustrator: Valerie Felts

Special thanks to Thomas M. Green for generating and
supplying the factorization charts at the back of this book.

ISBN 0-86651-304-3

Order Number DS01604   9 10 11-ML-99 98

DALE
SEYMOUR
PUBLICATIONS
P.O. BOX 10888
PALO ALTO, CA 94303

CONTENTS

INTRODUCTION

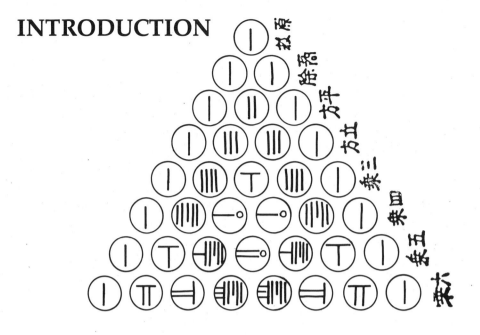

Pascal's Triangle, similar to the one shown above, was shown on a Japanese mural in the 1780s. The Chinese were aware of the Triangle in 1303.

Pascal's Triangle is a triangular array of numbers that has numerous applications in mathematics. The triangle is named for the great French mathematician Blaise Pascal (1623–1662); however, there is evidence that a famous Chinese algebraist, Chu Shi-kie, was familiar with the number pattern as early as 1303. Pascal is given credit for the triangle because of his extensive work with it and his application of it to the study of probability. Pascal and Pierre de Fermat (1601?–1665) are credited with the creation of probability theory.

Pascal discovered the triangle while exploring a problem presented to him by a gambler. Numbers in Pascal's Triangle reveal the number of ways in which a prescribed thing can be done or in which an event can happen. This is the basis for *combinatorial analysis*.

While the basic number pattern that makes up the triangle is simple, numerous other number patterns can be found that relate to various topics in arithmetic, algebra, geometry, and other branches of mathematics. This book, the first of two publications on patterns in Pascal's Triangle, emphasizes the visual patterns that are found in the rows and diagonals of the numerical array. Mathematics is often described as "the study of patterns." The incredible patterns that are introduced here are but a few that wait to be discovered by the reader who wishes to explore mathematics through the triangle.

D.G.S.

"THE BEST WAY TO LEARN ANYTHING IS TO DISCOVER IT BY YOURSELF."

George Polya

SUGGESTIONS FOR USING THIS BOOK

Mathematical patterns and concepts are more interesting and better remembered when they are personally discovered. It is strongly recommended that students be given the opportunity to discover the patterns or properties themselves rather than being shown the patterns that occur in Pascal's Triangle.

Classroom Use

It is suggested that students first be introduced to the triangle and the generation of the elements in the triangle. Once students understand these concepts, they can be presented with problems or explorations that are appropriate to their grade level, ability, and experience. Students should be given sufficient time to find the patterns. Tell them that when they discover answers to problems, they should keep these to themselves, thereby not depriving fellow students of the opportunity to discover the answers.

All worksheets and triangular grids can be reproduced from the copy masters in this book. Resource sheets with complete factorizations of the numbers in each row are available in the back of the book. These sheets can be helpful if students are searching for patterns that involve many rows in the triangle. If you are using an activity to develop or practice students' factoring skills, you might have the students work in teams, dividing up the work, looking for patterns, checking each other. When students are expected to generate *many* rows or *many* divisions, the use of calculators is highly recommended.

Visual Patterns Project

This project is designed to stimulate students' interest in looking for patterns in mathematics. Give each student a copy of the Visual Patterns Project worksheet, hexagonal grids, and, if necessary, factorization charts. *After* students have completed at least one of the patterns, the Visual Patterns problem set can be used to supplement the project. Hand out the Visual Patterns resource sheet, which includes all the patterns, with the problem set.

Bulletin Boards

After students have completed the worksheets, the two-color solution patterns can be used to create striking bulletin board displays. Hexagonal floor or counter tiles can be used to make permanent displays of the factoring patterns.

Pascal's Triangle

The companion to this book, *Pascal's Triangle* by Thomas M. Green and Charles L. Hamberg, contains more than 200 pages of additional information, problems, and exploration about the triangle. Most problems in *Pascal's Triangle* require a knowledge of first-year algebra.

STUDENT WORKSHEETS

PASCAL'S TRIANGLE DEFINED

Pascal's Triangle is a triangular array of natural numbers. The array is symmetrical about a vertical line from its top number. The sum of two adjacent numbers is equal to the number directly below and between them. The triangle continues infinitely.

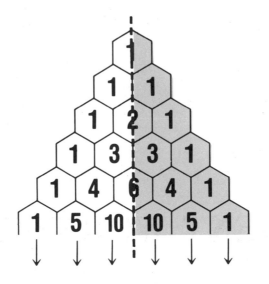

Numbers in a horizontal line make up *rows.*

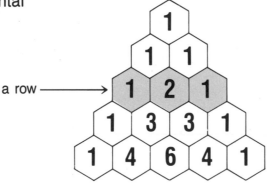

a row ⟶

Rows are numbered as follows:

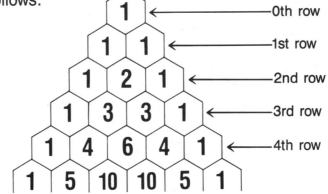

0th row
1st row
2nd row
3rd row
4th row

Numbers in an oblique line on the diagonal are called *diagonals*, or *columns*.

Diagonals are numbered as follows:

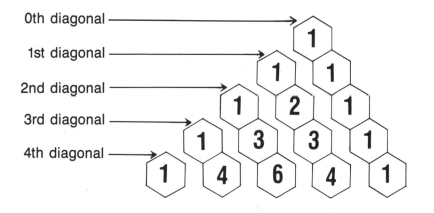

0th diagonal ⟶

1st diagonal ⟶

2nd diagonal ⟶

3rd diagonal ⟶

4th diagonal ⟶

A number is referred to as an *element*.

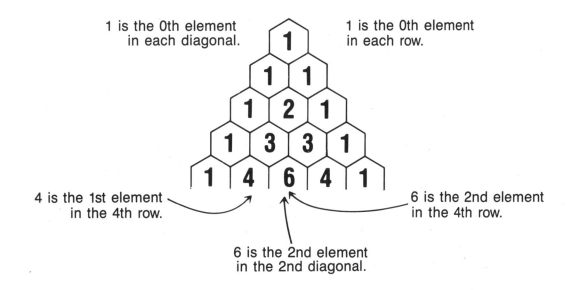

1 is the 0th element in each diagonal.

1 is the 0th element in each row.

4 is the 1st element in the 4th row.

6 is the 2nd element in the 4th row.

6 is the 2nd element in the 2nd diagonal.

The number of elements in a row is always one more than the number of the row. (For example, there are 5 elements in the 4th row.) The number of elements in a diagonal is infinite because the diagonal never ends.

PASCAL'S TRIANGLE

Use the pattern to fill in missing numbers in Pascal's Triangle.

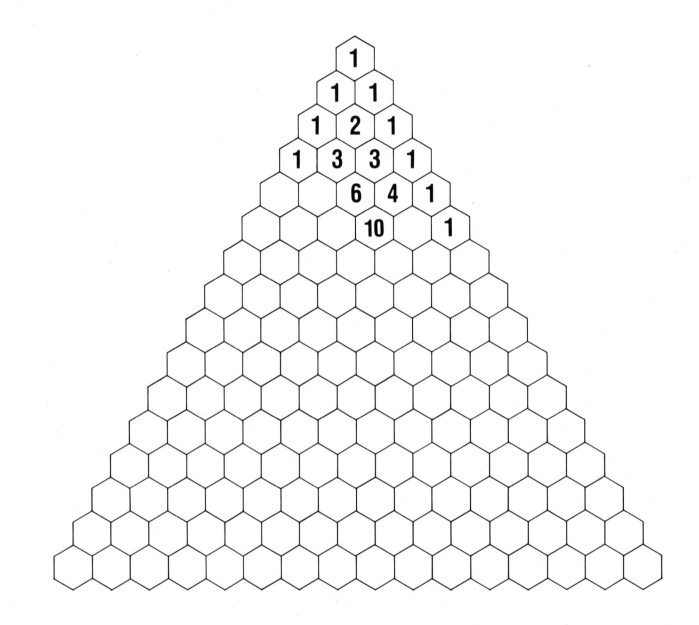

MISSING NUMBERS

Shown below are portions of Pascal's Triangle. Fill in the missing numbers.

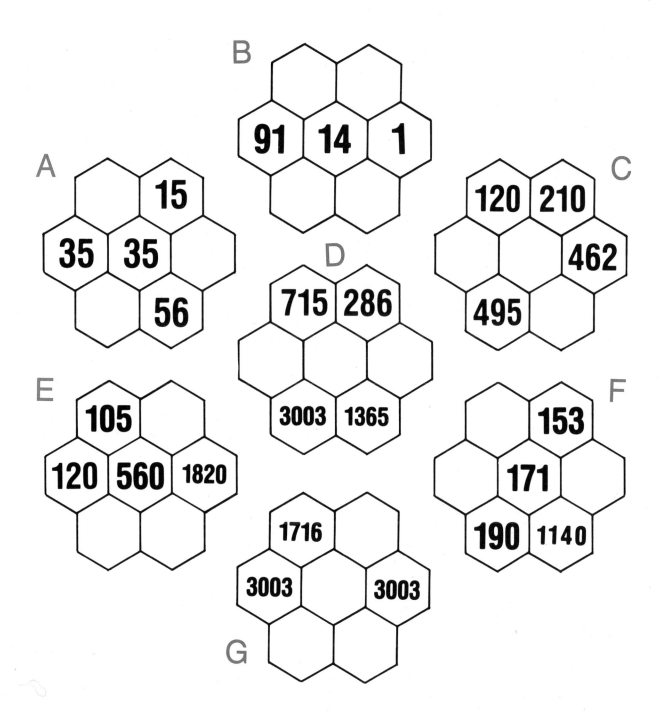

WHAT'S MY PATTERN?

Each of the triangular arrays of numbers shown below have been created by using different number patterns. See if you can discover each pattern and fill in the missing numbers.

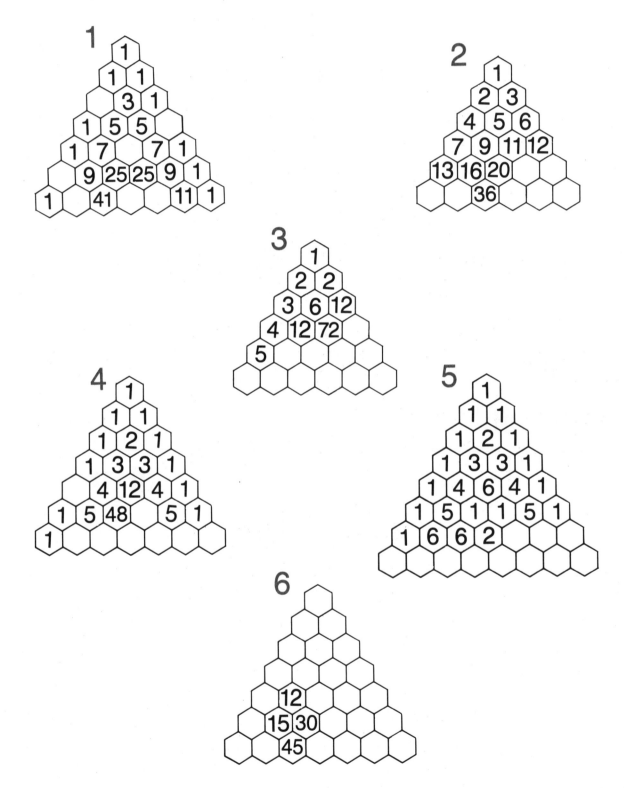

SUM PATTERNS

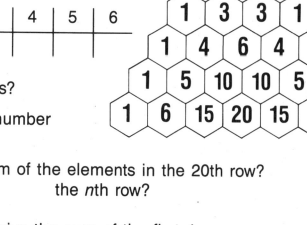

1. (a) Find the sum of the elements in the first few rows of Pascal's Triangle. Fill in the table below:

Row	0	1	2	3	4	5	6
Row sum	1	2					

   (b) What is the pattern of the sums?

   (c) How could you relate the row number to the sum of that row?

   (d) How would you express the sum of the elements in the 20th row? the 100th row? the nth row?

2. (a) Where is the element that will give the sum of the first 4 elements of the first diagonal $(1 + 2 + 3 + 4)$? the first 5 elements of the first diagonal?

   (b) Where is the element that will give the sum of the first 4 elements of the second diagonal $(1 + 3 + 6 + 10)$?

   (c) What is the pattern that will give the sum of any number of elements in any diagonal?

3. (a) Find the sum of *all* the elements in Pascal's Triangle down to and including the first 6 rows. Fill in the table below:

Row	0	1	2	3	4	5
Triangular sum	1	3				

   (b) If you see a pattern, then you can fill in the following table without adding all the elements.

Row	6	7	8	9	10
Triangular sum					

   (c) What is the rule?

FINDING PATHS

1. Use the grid below. Start at point A and only go right or up. How many routes are there from:

   (a) A to B?   (e) A to F?
   (b) A to C?   (f) A to G?
   (c) A to D?   (g) A to H?
   (d) A to E?   (h) A to I?

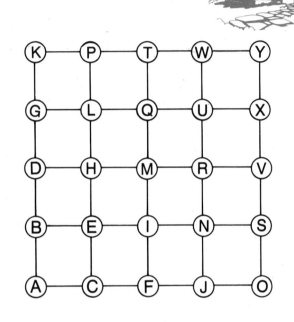

2. Write the number of routes from A to another letter next to that letter on the grid. Do you see a familiar pattern? How is Pascal's Triangle involved?

3. How many different routes are there from Home (A) to School (Y) (only moving right or up)?

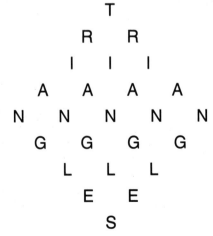

4. The solutions above may help you solve the following questions.
   (a) How many ways can you read the word "PASCALS"? (Read from top down.)

   (b) How many ways can you read the word "TRIANGLES"?

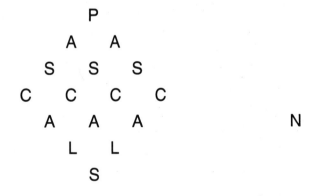

```
        P                                    T
      A   A                                R   R
    S   S   S                            I   I   I
  C   C   C   C                        A   A   A   A
    A   A   A                        N   N   N   N   N
      L   L                            G   G   G   G
        S                                L   L   L
                                           E   E
                                             S
```

KING STRUT'S CUBES

King Strut kept his gold in cubes. He enjoyed handling his gold and often spent time stacking these cubes. The king only stacked cubes directly on top of each other or side by side (never one behind the other).

The diagram below shows the different ways of stacking four gold cubes.

The Different Ways of Stacking 4 Gold Cubes

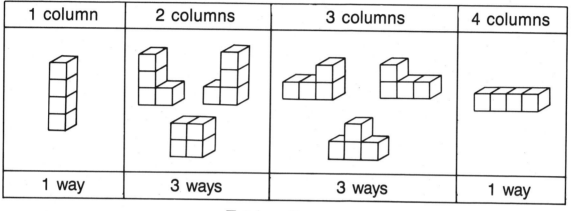

1 column	2 columns	3 columns	4 columns
1 way	3 ways	3 ways	1 way

Total = 8 ways

1. Make a diagram showing how three gold cubes could be stacked according to the king's rules.

2. Make a diagram showing how five gold cubes could be stacked.

3. Predict the number of ways six gold cubes could be stacked.

4. What is the pattern?

INTRODUCTION TO FIGURATE NUMBERS IN PASCAL'S TRIANGLE

The ancient Greeks were very interested in numbers that are connected by geometric forms. *Figurate numbers* (also known as polygonal numbers) are those numbers that can be arranged in a polygonal array.

Triangular numbers can be represented by dots in a triangular array.

Square numbers can be arranged in a square array and are of the form n^2 where n equals the number of dots on a side.

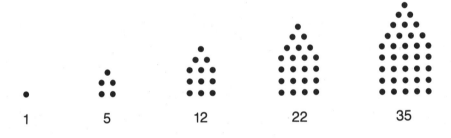

Pentagonal numbers can be arranged in a pentagonal array.

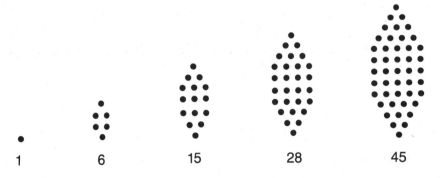

Hexagonal numbers can be shown as dots in an hexagonal array.

1. (a) What is the relationship between triangular numbers and square numbers?

   (b) Triangular and pentagonal numbers?

   (c) Triangular and hexagonal numbers?

POLYGONAL PATTERNS

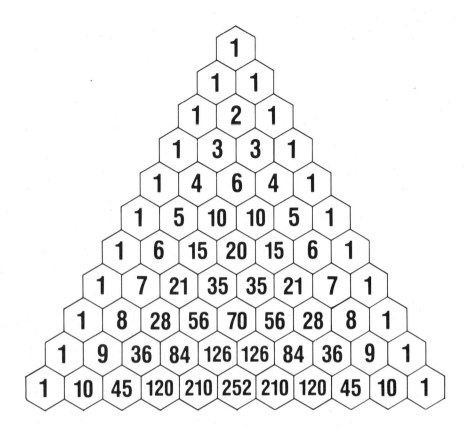

1. The square numbers 1, 4, 9, 16, 25, 36, . . . can be found in at least two patterns in Pascal's Triangle. What are these patterns? (Hint: The patterns involve the sum of certain numbers in the triangle.)

2. Can you find a pattern for pentagonal numbers? If so, what is it?

3. Look for the patterns for triangular numbers and hexagonal numbers. What do the patterns have in common?

HIDDEN PATTERNS

The problems below require you to search Pascal's Triangle for number patterns. These problems are meant to be challenging, so you may not find the answers right away. Explore, experiment. If you can't find a pattern, leave it and try another. Sometimes when you come back to a problem with a "fresh mind," you will think of a new approach that will reveal the answer.

1. (a) Look for the powers of 11 in the triangle. Where are they found?

   (b) How can 11^5 be explained by "regrouping"?

   (c) Using the same process explained in part (b), show the steps you would use to get 11^6.

2. The sequence 1, 1, 2, 3, 5, 8, 13, 21, 34, . . . is called the *Fibonacci sequence.* Each new element can be found by adding the two previous elements. These elements can be found in Pascal's Triangle. The pattern contains sets of numbers that when summed equal the elements of the Fibonacci sequence. What is the pattern? (Note: The pattern will reveal the sequence *in order.*)

3. (a) The six numbers that form a circle around another number in Pascal's Triangle are called a *ring.* The product of the six numbers in a ring has a special property. What is that property?

   (b) Give some examples.

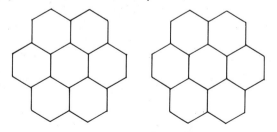

INTRODUCTION TO COMBINATIONS

Combinations of things are important in mathematics. It is often necessary to list all possible combinations that could occur in a given situation. For instance, suppose you wanted three records but only had enough money for two. There are three possible combinations of records you could purchase—*a/b, a/c,* or *b/c.* It is important to remember that when listing combinations, the elements of the set are important but the way they are arranged is not. That is, *a/b* and *b/a* are not considered different combinations.

1. Sue has four items in her lunch: an apple, a banana, a sandwich, and a glass of milk. She may eat none of the items, all of the items, or any combination of the items. Use the table below to list all of the choices Sue has. [Use initials to save time and space—apple (A), banana (B), milk (M), and sandwich (S).]

Number of choices	Combinations of items eaten
1	If Sue eats 0 items
4	If Sue eats 1 item
6	If Sue eats 2 items
4	If Sue eats 3 items
1	If Sue eats 4 items

Total Possible
Choices 16

2. Suppose Sue also has peanuts in her lunch. Make a table similar to the one above showing all the combinations and number of choices Sue has with five items.

3. Predict the *total* number of choices Sue would have if she had 10 items in her lunch.   15 items.

4. How are the combinations related to Pascal's Triangle?

COMBINATIONS FORMULA

A combination is any collection of *r* members chosen out of a set of *n* things or a *combination of **n** things taken **r** at a time.* For example, suppose you had a set of three objects (a cat, a dog, and a bird) and you wanted to combine them in groups of two (cat/dog, cat/bird, dog/bird). You would say that this is a "combination of 3 things taken 2 at a time."

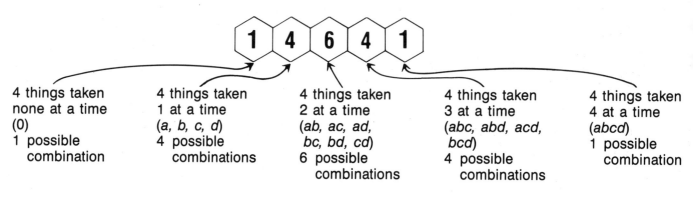

| 4 things taken none at a time (0) 1 possible combination | 4 things taken 1 at a time (*a, b, c, d*) 4 possible combinations | 4 things taken 2 at a time (*ab, ac, ad, bc, bd, cd*) 6 possible combinations | 4 things taken 3 at a time (*abc, abd, acd, bcd*) 4 possible combinations | 4 things taken 4 at a time (*abcd*) 1 possible combination |

Another way to write combinations is as follows:

$\binom{4}{2}$ means the combination of 4 things taken 2 at a time.

$\binom{8}{3}$ means the combination of 8 things taken 3 at a time.

$\binom{n}{r}$ means the combination of *n* things taken *r* at a time.

Therefore, you could represent the elements in the fourth row as follows:

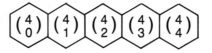

The general formula for *n* things taken *r* at a time is $\dfrac{n!}{(n-r)!\,r!}$

where *n*! is read as "*n* factorial." The factorial of a positive *n* is $1 \cdot 2 \cdot 3 \cdot 4 \cdot \cdots \cdot (n-1) \cdot n$. For example, $5! = 1 \cdot 2 \cdot 3 \cdot 4 \cdot 5 = 120$. $4! = 1 \cdot 2 \cdot 3 \cdot 4 = 24$. 0! is equal to 1.

FINDING COMBINATIONS

Use the formula $\binom{n}{r} = \dfrac{n!}{(n-r)!\,r!}$ to find the following:

1. $\binom{6}{2} = \dfrac{6!}{(6-2)!\,2!} = \dfrac{6 \cdot 5 \cdot 4 \cdot 3 \cdot 2 \cdot 1}{4! \cdot 2!} = \dfrac{6 \cdot 5 \cdot 4 \cdot 3 \cdot 2 \cdot 1}{(4 \cdot 3 \cdot 2 \cdot 1) \cdot (2 \cdot 1)}$

$= \dfrac{6 \cdot 5 \cdot 4 \cdot 3 \cdot 2 \cdot 1}{4 \cdot 3 \cdot 2 \cdot 1 \cdot 2 \cdot 1} = 15$

2. $\binom{10}{3} =$

3. $\binom{5}{5} =$

4. $\binom{4}{1} =$

5. $\binom{20}{4} =$

The eighth and ninth rows of Pascal's Triangle are:

1	8	28	56	70	56	28	8	1
1	9	36	84	126	126	84	36	9	1

These can be rewritten as:

$$\binom{8}{0}\ \binom{8}{1}\ \binom{8}{2}\ \binom{8}{3}\ \binom{8}{4}\ \binom{8}{5}\ \binom{8}{6}\ \binom{8}{7}\ \binom{8}{8}$$

$$\binom{9}{0}\ \binom{9}{1}\ \binom{9}{2}\ \binom{9}{3}\ \binom{9}{4}\ \binom{9}{5}\ \binom{9}{6}\ \binom{9}{7}\ \binom{9}{8}\ \binom{9}{9}$$

Use the formula $\binom{n}{r} = \dfrac{n!}{(n-r)!\,r!}$ to find the following: (A calculator will help.)

1. The combination of 7 things taken 4 at a time.

2. $\binom{10}{4}$

3. The third number in the eleventh row of Pascal's Triangle.

4. The fourth number in the fourteenth row of Pascal's Triangle.

5. $\binom{22}{19}$

BINOMIALS (algebra is a prerequisite)

You may recall from your study of algebra that a *binomial* is an expression that involves the addition or subtraction of two terms. Some examples of binomials are:

$3x + y \qquad x + 5 \qquad x^2 - 4 \qquad a + b$

The general form of a binomial is $(a + b)$. If we square $(a + b)$, we get $(a + b)^2$. We can *expand* the expression to $a^2 + 2ab + b^2$. The steps for this expansion are:

$$
\begin{array}{r}
a + b \\
\times\ a + b \\
\hline
+ab + b^2 \\
a^2 + ab \\
\hline
a^2 + 2ab + b^2
\end{array}
$$

Two is the *numerical coefficient* of ab. The numerical coefficient of both a^2 and b^2 is understood to be 1. The 1 is usually not written.

1. Expand the following expressions:

   $(a + b)^0 = 1$

   $(a + b)^1 = 1a + 1b$

   $(a + b)^2 = 1a^2 + 2ab + 1b^2$

   $(a + b)^3 =$

   $(a + b)^4 =$

   $(a + b)^5 =$

   $(a + b)^6 =$

Optional:

   $(a + b)^n =$

2. What is the relationship between the binomial expansion and Pascal's Triangle?

VISUAL PATTERNS PROJECT

Mathematics is often described as the *study of patterns.* Learning to look for patterns in all aspects of mathematics will help make the work more entertaining and understandable. The purpose of this project is to discover some of the striking *visual* patterns in Pascal's Triangle.

Instructions:

Materials: Hexagonal grid triangles (if larger triangles are needed, make copies of the grids and carefully tape them together) felt-tip pen or marker calculator or factorization reference charts

Choose a whole number between 1 and 11. Color in all of the elements in Pascal's Triangle that are divisible by that number. (Using only one color will make the patterns easier to see.) Do as many rows as necessary to determine the geometrical pattern. If you carry the design past 20 rows, you may wish to use a calculator or factorization charts to assist you in factoring the elements. (Numbers near the center of the rows become very large after the 30th row.)

Example

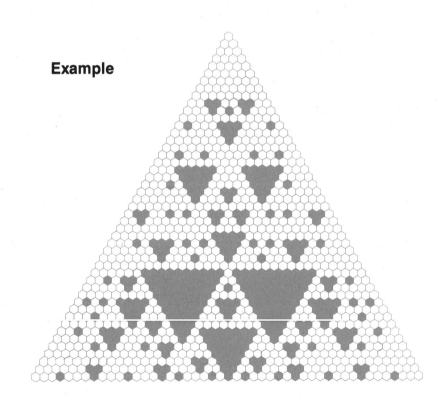

VISUAL PATTERNS PROBLEM SET

1. Which multiples of the numbers 2 through 10 make patterns that have mirror symmetry at *each* of the three vertices of the triangle?

2. Look at the pattern for the multiples of 3. Would you expect one large triangle or two large triangles to begin in row 54? Why? How could you test this?

3. What is the general visual pattern for the multiples of each *prime* number in Pascal's Triangle?

4. Look at the pattern for the multiples of 4.
   (a) What is true about rows numbered $2^1 - 1$, $2^2 - 1$, $2^3 - 1$, $2^4 - 1$, $2^5 - 1$, $2^n - 1$?

   (b) What other rows have this property?

   (c) What is their pattern?

5. (a) What happens to the percentage of the elements that are multiples of a given number in Pascal's Triangle as the number of rows plotted increases?

   (b) Make a *guess* as to what percent of the numbers in Pascal's Triangle are divisible by 7 in the first 49 rows. In the first 343 rows. In the first million rows.

To do these problems you need to use the patterns on page 19.

VISUAL PATTERNS RESOURCE PAGE

Multiples of 2

Multiples of 3

Multiples of 4

Multiples of 5

Multiples of 6

Multiples of 7

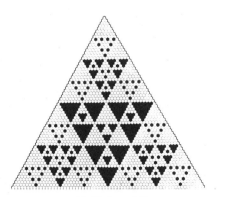

Multiples of 8

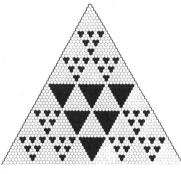

Multiples of 9

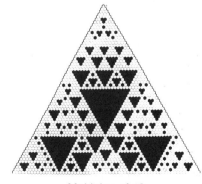

Multiples of 10

PROBLEM SOLUTIONS
AND
BULLETIN BOARD
DISPLAY SHEETS

Pascal's Triangle

See Chart 1 on page 29.

Missing Numbers

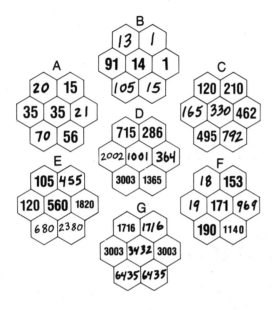

What's My Pattern?

1

2

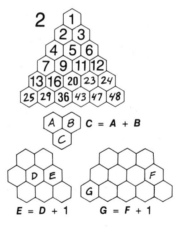

3

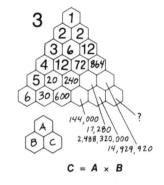

$C = A \times B$

4

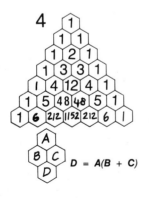

5

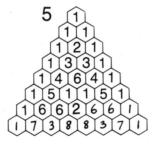

Each element equals the sum of the digits in the equivalent element on Pascal's Triangle. [10 (original) = 1 + 0 = 1 (new)]

6

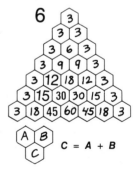

Sum Patterns

1. (a)

Row	0	1	2	3	4	5	6
Row sum	1	2	4	8	16	32	64

(b) The sums double each time. (See Chart 2 on page 31.)
(c) The sum of a row equals 2 to the power of the row.
(d) 2^{20}, 2^{100}, 2^n

2. (a) The 4th element of the second diagonal; the 5th element of the second diagonal
(b) The 4th element of the third diagonal
(c) The nth in the $(r + 1)$ diagonal, where r is the number of the diagonal containing the summed elements. (See Chart 3 on page 33.)

3. (a)

Row	0	1	2	3	4	5
Triangular sum	1	3	7	15	31	63

(b)

Row	6	7	8	9	10
Triangular sum	$2^7 - 1$	$2^8 - 1$	$2^9 - 1$	$2^{10} - 1$	$2^{11} - 1$

(c) Raise 2 to the power one greater than the row, then subtract 1.
$2^{n+1} - 1$, where n is the number of the row.

Finding Paths

1. (a) 1   (e) 1
   (b) 1   (f) 1
   (c) 1   (g) 3
   (d) 2   (h) 3

2. (a) The pattern is Pascal's Triangle. The number of possible routes gives the elements of the triangle.
   (b) 70

3. (a) 20   (b) 70

King Strut's Cubes

1.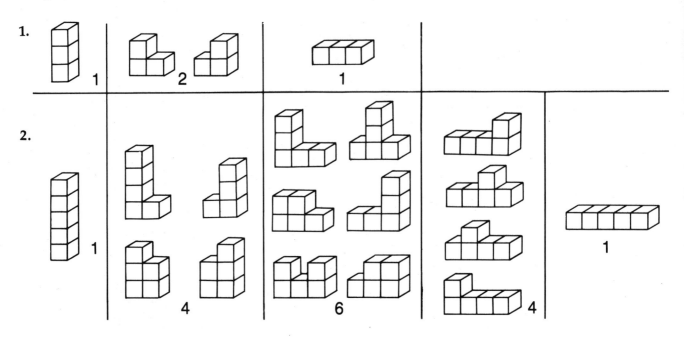

3. 1 way for 1 column, 5 ways for 2 columns, 10 ways for 3 columns, 10 ways for 4 columns, 5 ways for 5 columns, 1 way for 6 columns

4. The number of ways the cubes could be stacked makes up Pascal's Triangle.

Introduction to Figurate Numbers

1. (a) Square numbers are the sum of two consecutive triangular numbers.
   Square number x = Triangular number $(x - 1)$ + Triangular number x

   (b) Pentagonal numbers are the sum of a square number and a triangular number (or at least three triangular numbers).
   Pentagonal number x = Square number x + Triangular number $(x - 1)$

   (c) Hexagonal numbers are the sum of two triangular numbers and a square number.
   Hexagonal number x = Square number x + 2 [Triangular number $(x - 1)$]

Polygonal Patterns

1. See Charts 4 and 5 on pages 35 and 37.

2. See Chart 6 on page 39.

3. The elements of both triangular numbers and hexagonal numbers can be found in the 2nd diagonal of Pascal's Triangle. The triangular numbers correspond to each element of the diagonal, and the hexagonal numbers correspond to every other element of the diagonal (starting with the 0th element).

Hidden Patterns

1. (a) See Chart 7 on page 41.
   (b) See Chart 7.
   (c) See Chart 7.

2. See Chart 8 on page 43.

3. (a) The products of rings are square numbers. (See Chart 9 on page 45.)

(b)
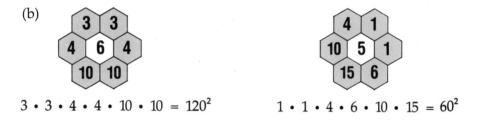

$3 \cdot 3 \cdot 4 \cdot 4 \cdot 10 \cdot 10 = 120^2$ $1 \cdot 1 \cdot 4 \cdot 6 \cdot 10 \cdot 15 = 60^2$

Introduction to Combinations

1. 0; A, B, M, S; AB, AM, AS, BM, BS, MS; ABM, ABS, AMS, BMS; ABMS

2.

Number of choices	Combinations of items eaten
1	0
5	A, B, M, S, N
10	AB, AM, AS, AN, BM, BS, BN, MS, MN, NS
10	ABM, ABS, ABN, ANM, ANS, AMS, BNM, BNS, BMS, NMS
5	ABNM, ABNS, ABMS, ANMS, BNMS
1	ABMNS

3. (a) 1054
   (b) 32768

4. Given any number of items, *n*, the number of combinations corresponds to the elements in the *n*th row. The total number of combinations equals the sum of all the elements in the row.

Finding Combinations

1. $\dfrac{6!}{(6-2)!\ 2!} = \dfrac{6 \cdot 5 \cdot 4 \cdot 3 \cdot 2 \cdot 1}{4!\ 2!} = \dfrac{6 \cdot 5 \cdot \cancel{4} \cdot \cancel{3} \cdot \cancel{2} \cdot \cancel{1}}{\cancel{4} \cdot \cancel{3} \cdot 2 \cdot 1 \cdot \cancel{2} \cdot \cancel{1}} = 15$

2. $\dbinom{10}{3} = \dfrac{10!}{(10-3)!\ 3!} = \dfrac{10!}{7!\ 3!} = \dfrac{10 \cdot 9 \cdot 8 \cdot \cancel{7} \cdot \cancel{6} \cdot \cancel{5} \cdot \cancel{4} \cdot \cancel{3} \cdot \cancel{2} \cdot \cancel{1}}{\cancel{7} \cdot \cancel{6} \cdot \cancel{5} \cdot \cancel{4} \cdot \cancel{3} \cdot \cancel{2} \cdot \cancel{1} \cdot 3 \cdot 2 \cdot 1} = 120$

3. $\dbinom{5}{5} = \dfrac{5!}{(0)!\ (5)!} = \dfrac{5!}{5!} = 1$

4. $\dbinom{4}{1} = \dfrac{4!}{(4-1)!\ 1!} = \dfrac{4!}{3!\ 1!} = \dfrac{4 \cdot \cancel{3} \cdot \cancel{2} \cdot \cancel{1}}{\cancel{3} \cdot \cancel{2} \cdot \cancel{1} \cdot 1} = 4$

5. $\dbinom{20}{4} = \dfrac{20!}{(20-4)!\ 4!} = \dfrac{20!}{16!\ 4!} = \dfrac{\cancel{20}^{5} \cdot 19 \cdot \cancel{18}^{3} \cdot 17}{\cancel{4} \cdot \cancel{3} \cdot \cancel{2} \cdot 1} = 4845$

1. $\dbinom{7}{4} = \dfrac{7!}{3!\ 4!} = \dfrac{7 \cdot \cancel{6} \cdot 5 \cdot \cancel{4} \cdot \cancel{3} \cdot \cancel{2} \cdot 1}{\cancel{3} \cdot \cancel{2} \cdot 1 \cdot \cancel{4} \cdot \cancel{3} \cdot \cancel{2} \cdot \cancel{1}} = 35$

2. $\dbinom{10}{4} = \dfrac{10!}{6!\ 4!} = \dfrac{10 \cdot \cancel{9}^{3} \cdot \cancel{8} \cdot 7 \cdot \cancel{6} \cdot \cancel{5} \cdot \cancel{4} \cdot \cancel{3} \cdot \cancel{2} \cdot \cancel{1}}{\cancel{6} \cdot \cancel{5} \cdot \cancel{4} \cdot \cancel{3} \cdot \cancel{2} \cdot \cancel{1} \cdot \cancel{4} \cdot \cancel{3} \cdot \cancel{2} \cdot 1} = 210$

3. $\dbinom{11}{2} = \dfrac{11!}{9!\ 2!} = \dfrac{11 \cdot 10}{2 \cdot 1} = 55$

4. $\dbinom{14}{3} = \dfrac{14!}{11!\ 3!} = \dfrac{14 \cdot 13 \cdot 12}{3 \cdot 2 \cdot 1} = 364$

5. $\dbinom{22}{19} = \dfrac{22!}{3!\ 19!} = \dfrac{22 \cdot 21 \cdot 20}{3 \cdot 2 \cdot 1} = 1540$

Binomials

1. See Chart 10 on page 47.

2. Numerical coefficients are the terms in the expansion of elements in Pascal's Triangle.

Visual Patterns Problem Sheet

1. Multiples of 2, 3, 5, 7, 8, and 9 have mirror symmetry at each of the three vertices.

2. Two large triangles will come next. This continues the pattern of groups of 3 triangles that are the same size. You could test this by checking to see if the middle number in the row is divisible by 3.

3. The pattern formed by multiples of prime numbers is a pattern of equilateral triangles whose edge length is always one less than the multiple being plotted. That is, if multiples of 5 were being plotted, the triangle would have an edge length of 4. These are called *first generation* triangles. A *second generation* triangle can be found below the first generation. The basic pattern of the second generation is identical to the first generation—if there were 10 triangular elements in the first generation, there will be 10 triangular elements in the second generation. However, the edge length of the individual elements is one less than the *square* of the multiple being plotted. (The second generation triangles for multiples of 5 would have an edge length of 24.) This pattern continues to infinity.

4. (a) None of these rows are plotted (they contain no elements that are divisible by 4).
   (b) Rows 11, 23, and 48 have this property.
   (c) The pattern is $4(3n) - 1$.

5. (a) The percentage increases.
   (b) First 7 rows: 0% are divisible by 7; first 48 rows: 36% are divisible by 7; first 343 rows: 63% are divisible by 7; first one million rows: over 99.99% are divisible by 7. (See Charts 11, 12, and 13 on pages 49, 51, and 53.)

CHART 1

PASCAL'S TRIANGLE

CHART 2

SUM OF A ROW
IN PASCAL'S TRIANGLE

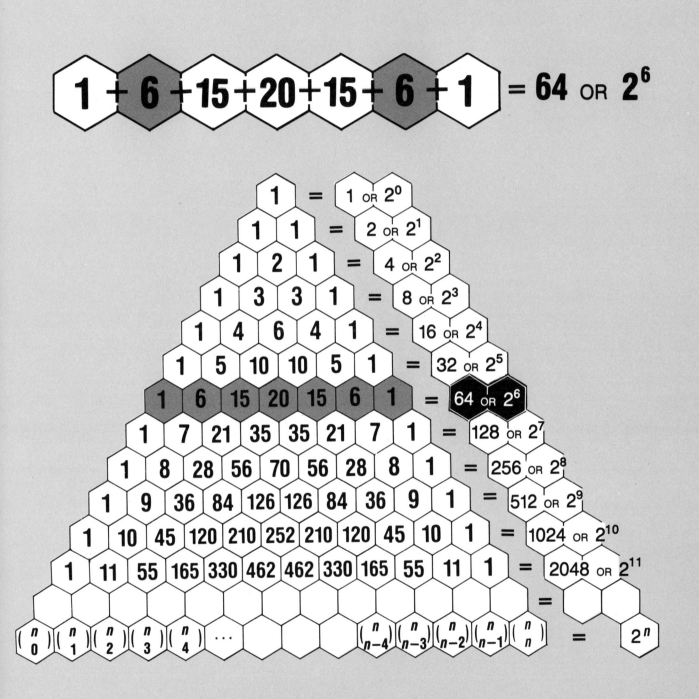

CHART 3

SUM OF A DIAGONAL
IN PASCAL'S TRIANGLE

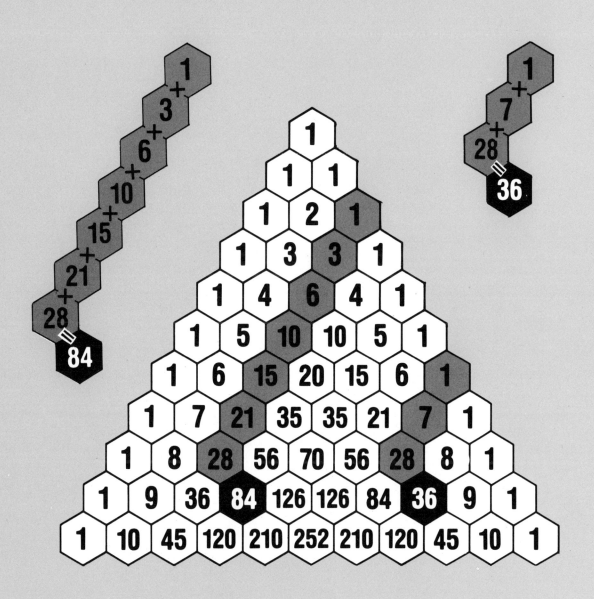

CHART 4

SQUARE NUMBERS
IN PASCAL'S TRIANGLE

Example 1

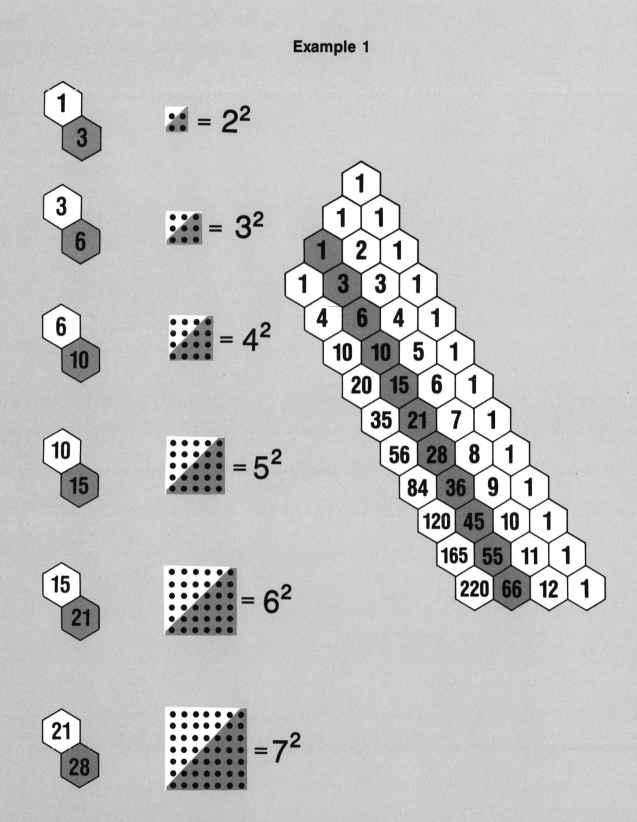

CHART 5

SQUARE NUMBERS
IN PASCAL'S TRIANGLE

Example 2

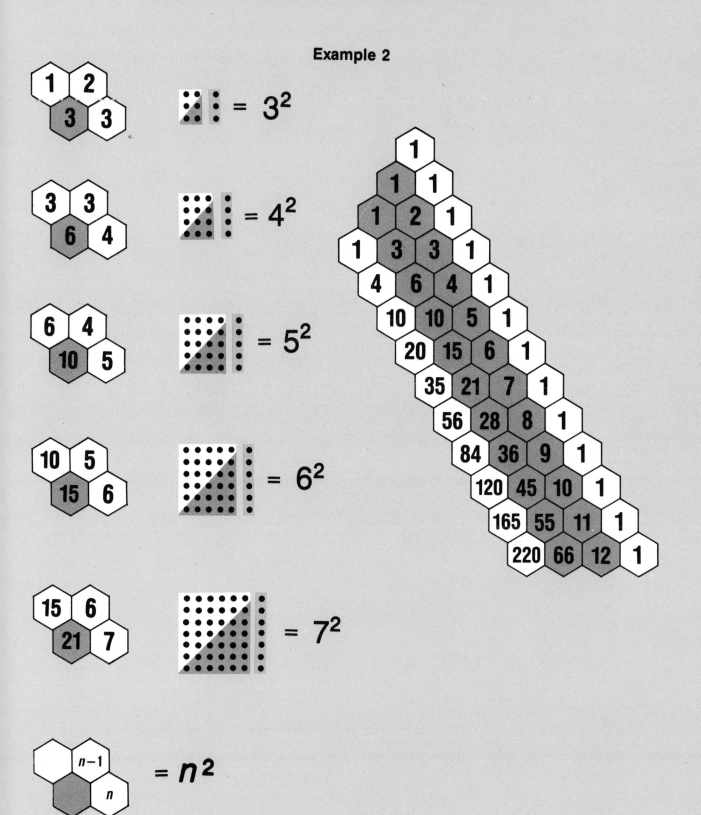

$$\boxed{\cdot} = 3^2$$

$$= 4^2$$

$$= 5^2$$

$$= 6^2$$

$$= 7^2$$

$$= n^2$$

CHART 6

PENTAGONAL NUMBERS
IN PASCAL'S TRIANGLE

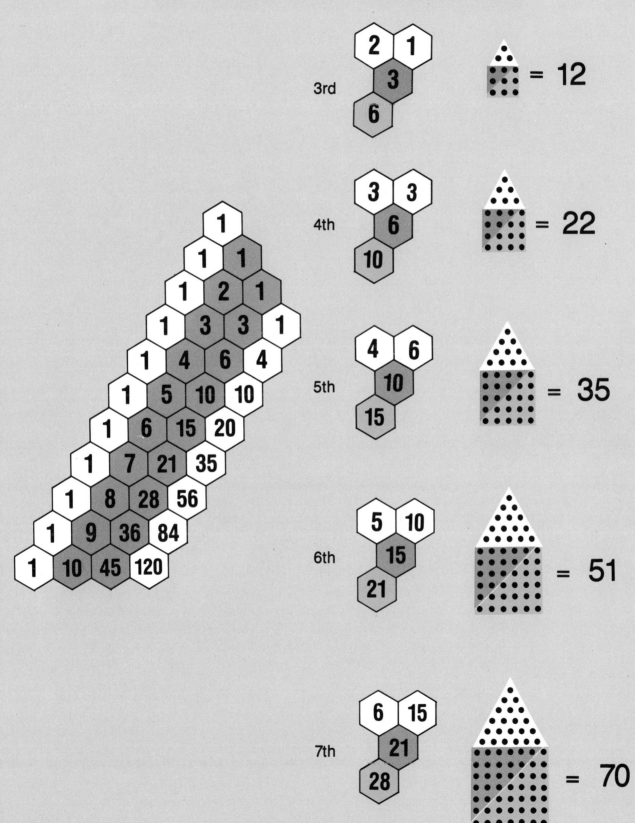

CHART 7

POWERS OF ELEVEN
IN PASCAL'S TRIANGLE

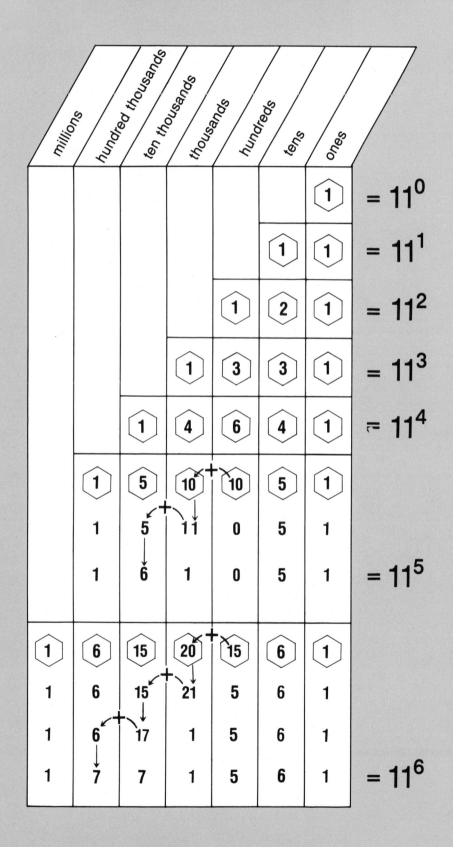

CHART 8

FIBONACCI NUMBERS IN PASCAL'S TRIANGLE

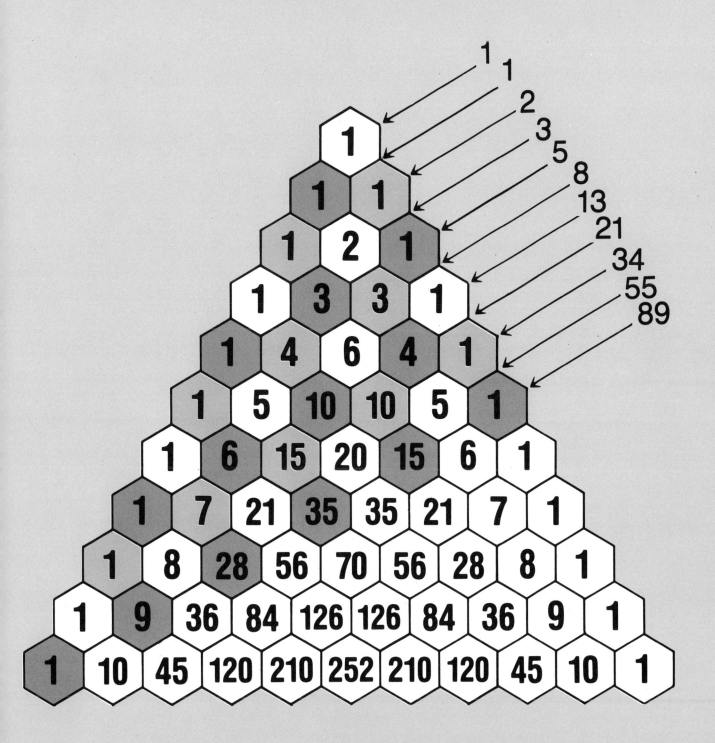

CHART 9

THE PRODUCTS OF RINGS
IN PASCAL'S TRIANGLE

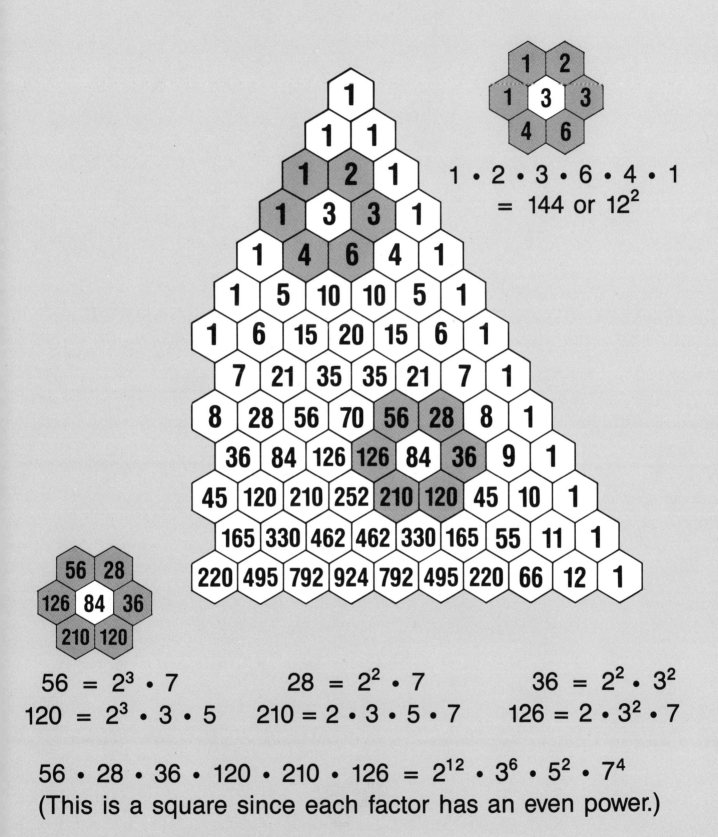

$1 \cdot 2 \cdot 3 \cdot 6 \cdot 4 \cdot 1$
$= 144$ or 12^2

$56 = 2^3 \cdot 7$ $28 = 2^2 \cdot 7$ $36 = 2^2 \cdot 3^2$

$120 = 2^3 \cdot 3 \cdot 5$ $210 = 2 \cdot 3 \cdot 5 \cdot 7$ $126 = 2 \cdot 3^2 \cdot 7$

$56 \cdot 28 \cdot 36 \cdot 120 \cdot 210 \cdot 126 = 2^{12} \cdot 3^6 \cdot 5^2 \cdot 7^4$

(This is a square since each factor has an even power.)

CHART 10

THE BINOMIAL EXPANSION AND PASCAL'S TRIANGLE

$(a + b)^0 = 1$

$(a + b)^1 = 1a + 1b$

$(a + b)^2 = 1a^2 + 2ab + 1b^2$

$(a + b)^3 = 1a^3 + 3a^2b + 3ab^2 + 1b^3$

$(a + b)^4 = 1a^4 + 4a^3b + 6a^2b^2 + 4ab^3 + 1b^4$

$(a + b)^5 = 1a^5 + 5a^4b + 10a^3b^2 + 10a^2b^3 + 5ab^4 + 1b^5$

$(a + b)^6 = 1a^6 + 6a^5b + 15a^4b^2 + 20a^3b^3 + 15a^2b^4 + 6ab^5 + 1b^6$

$\cdots$

$(a + b)^n = \binom{n}{0}a^n + \binom{n}{1}a^{n-1}b + \binom{n}{2}a^{n-2}b^2 + \binom{n}{3}a^{n-3}b^3 +$

$\cdots + \binom{n}{n-2}a^2b^{n-2} + \binom{n}{n-1}ab^{n-1} + \binom{n}{n}b^n$

CHART 11

MULTIPLES OF SEVEN
IN PASCAL'S TRIANGLE

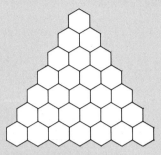

Generation 1
49 rows*
1225 numbers
36% multiples of 7

Generation 0
7 rows*
28 numbers
0% multiples of 7

Generation 2
343 rows*
58,996 numbers
63% multiples of 7
(5.1 ft. edge length
at scale of generation 0)

*Counted here as actual rows as opposed to numbering system in
Pascal's Triangle, which counts the first row as row 0.

CHART 12

MULTIPLES OF SEVEN IN PASCAL'S TRIANGLE

Generation 3
2401 rows*
2,883,601 numbers
79% multiples of 7
(36 ft edge length
at scale of generation 0)

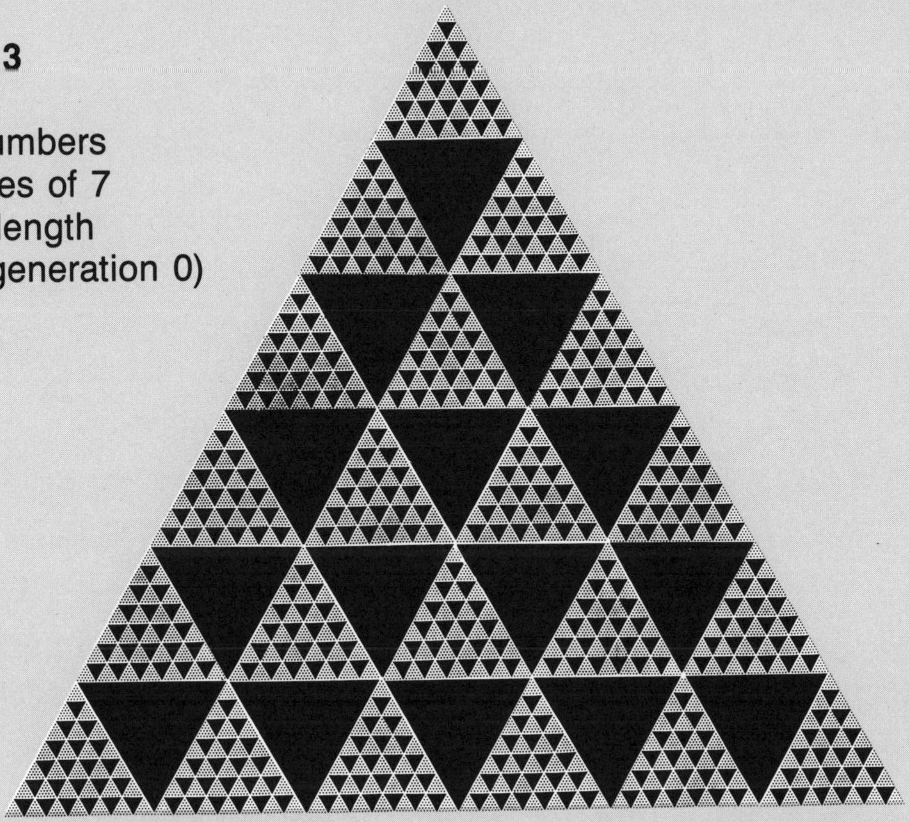

Generation 4 would have 16,807 rows*; 141,246,028 numbers; 88% multiples of 7; (252 ft edge length at scale of generation 0).

Generation 5 would have 117,649 rows*; 6,920,702,425 numbers; 93% multiples of 7; (1765 ft edge length at scale of generation 0).

Generation 6 would have 823,543 rows*; 1.1980 x 10^{12} numbers; 99.9% multiples of 7; (2.3 mi edge length at scale of generation 0).

*Counted here as actual rows as opposed to numbering system in Pascal's Triangle, which counts the first row as row 0.

CHART 13

Cells Divisible by Seven in Pascal's Triangle

Number of Cells Divisible by Seven

Row*	Triangle size 1: 21 cells	Triangle size 2: 1176 cells	Triangle size 3: 58,653 cells	Triangle size 4: 2,881,200 cells	Triangle size 5: 141,229,221 cells	Total cells divisible by 7	Groups of 28 cells not divisible by 7	Total cells not divisible by 7	Ratio of cells divisible by 7	Percent of cells divisible by 7
7 (7^1)	0	0	0	0	0	0	1	28	$\frac{0}{28}$	0%
49 (7^2)	21	0	0	0	0	441	28	784	$\frac{441}{1225}$	36%
343 (7^3)	$21 \cdot 28$	21	0	0	0	37,044	28^2	21,952	$\frac{37,044}{58,996}$	63%
2401 (7^4)	$21 \cdot 28^2$	$21 \cdot 28$	21	0	0	2,268,945	28^3	614,656	$\frac{2,268,945}{2,883,601}$	79%
16,807 (7^5)	$21 \cdot 28^3$	$21 \cdot 28^2$	$21 \cdot 28$	21	0	124,035,660	28^4	17,210,368	$\frac{124,035,660}{141,246,028}$	88%
117,649 (7^6)	$21 \cdot 28^4$	$21 \cdot 28^3$	$21 \cdot 28^2$	$21 \cdot 28$	21	6,438,812,121	28^5	481,890,304	$\frac{6,438,812,121}{6,920,702,425}$	93%
823,543 (7^7)	$21 \cdot 28^5$	$21 \cdot 28^4$	$21 \cdot 28^3$	$21 \cdot 28^2$	$21 \cdot 28$	1.1975×10^{12}	28^6	4.8189×10^8	$\frac{1.1975 \times 10^{12}}{1.1980 \times 10^{12}}$	99.9%

*All rows are counted here as actual rows as opposed to the numbering system in Pascal's Triangle, which counts the first row as 0

VISUAL PATTERNS
IN
PASCAL'S
TRIANGLE

VISUAL PATTERNS IN PASCAL'S TRIANGLE

Multiples of 2

VISUAL PATTERNS IN PASCAL'S TRIANGLE

Multiples of 3

VISUAL PATTERNS IN PASCAL'S TRIANGLE

Multiples of 4

61

VISUAL PATTERNS IN PASCAL'S TRIANGLE

Multiples of 5

VISUAL PATTERNS IN PASCAL'S TRIANGLE

Multiples of 6

VISUAL PATTERNS IN PASCAL'S TRIANGLE

Multiples of 7

VISUAL PATTERNS IN PASCAL'S TRIANGLE

Multiples of 8

VISUAL PATTERNS IN PASCAL'S TRIANGLE

Multiples of 9

VISUAL PATTERNS IN PASCAL'S TRIANGLE

Multiples of 10

BLANK GRID SHEETS

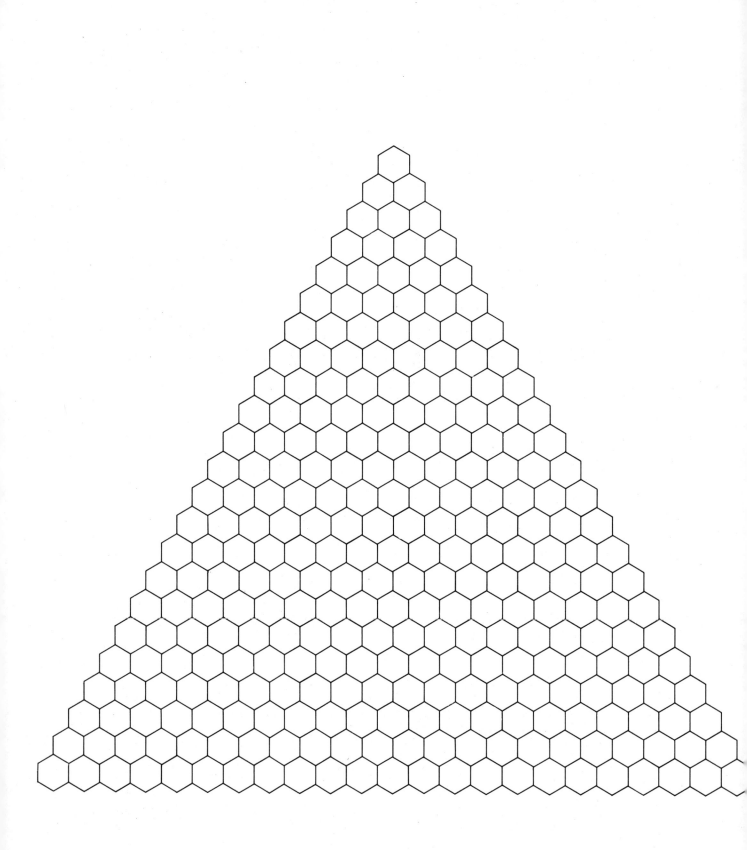

76

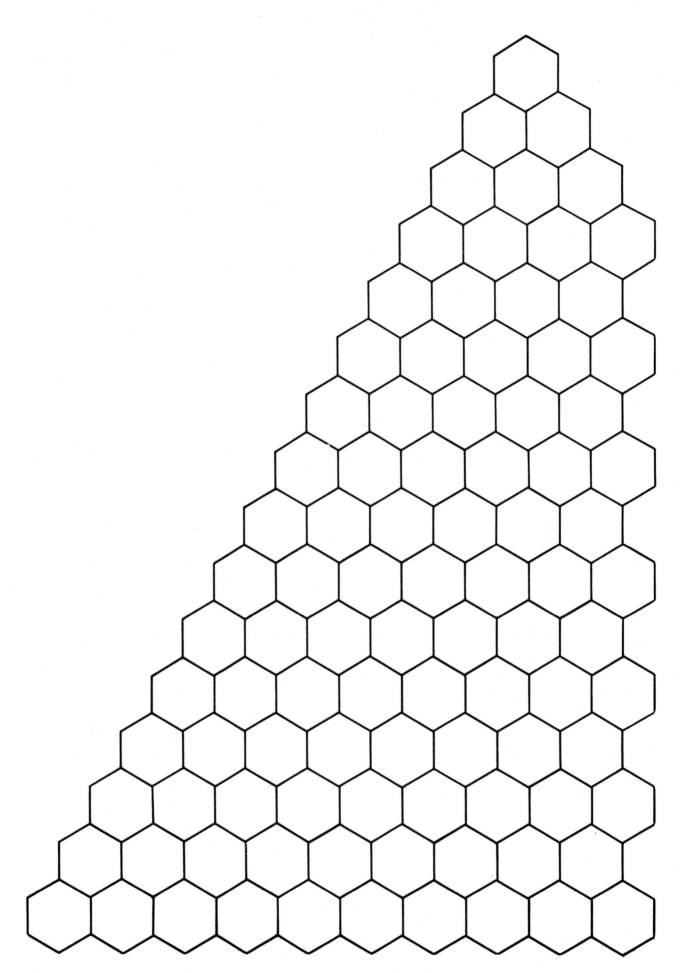

78

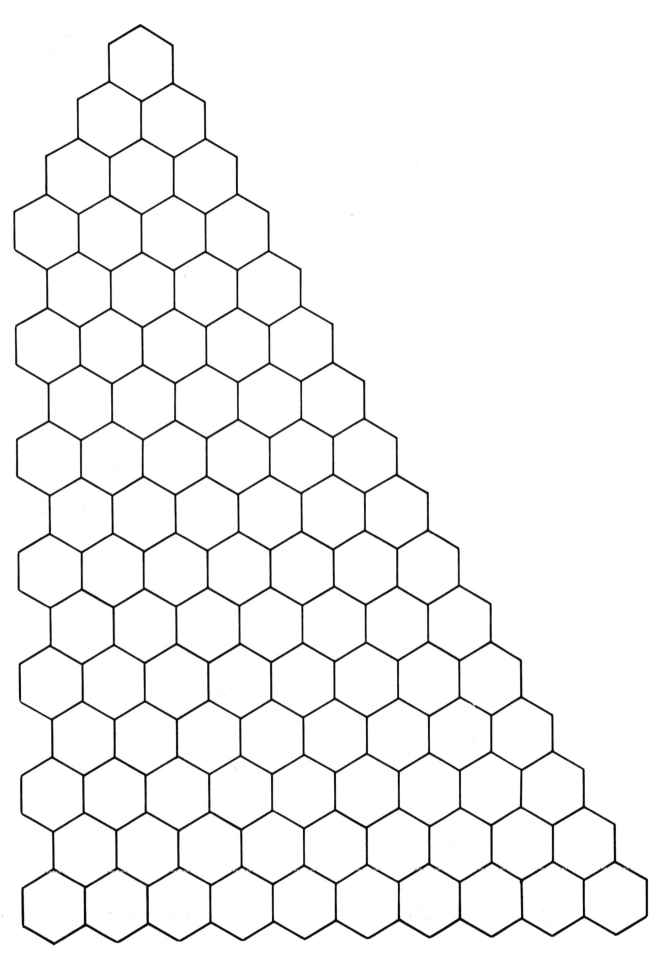

79

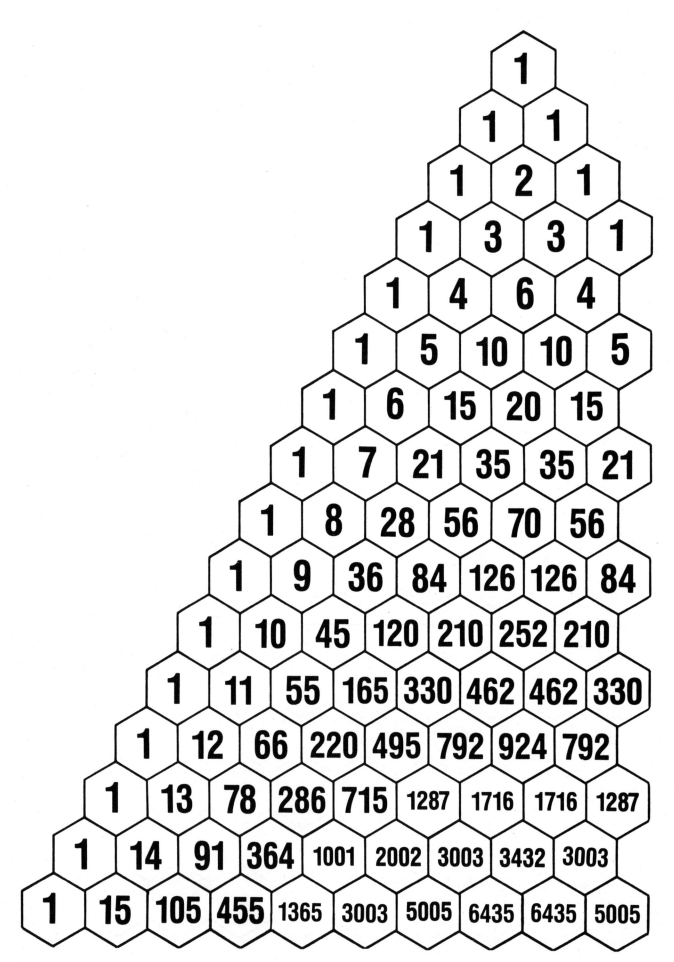

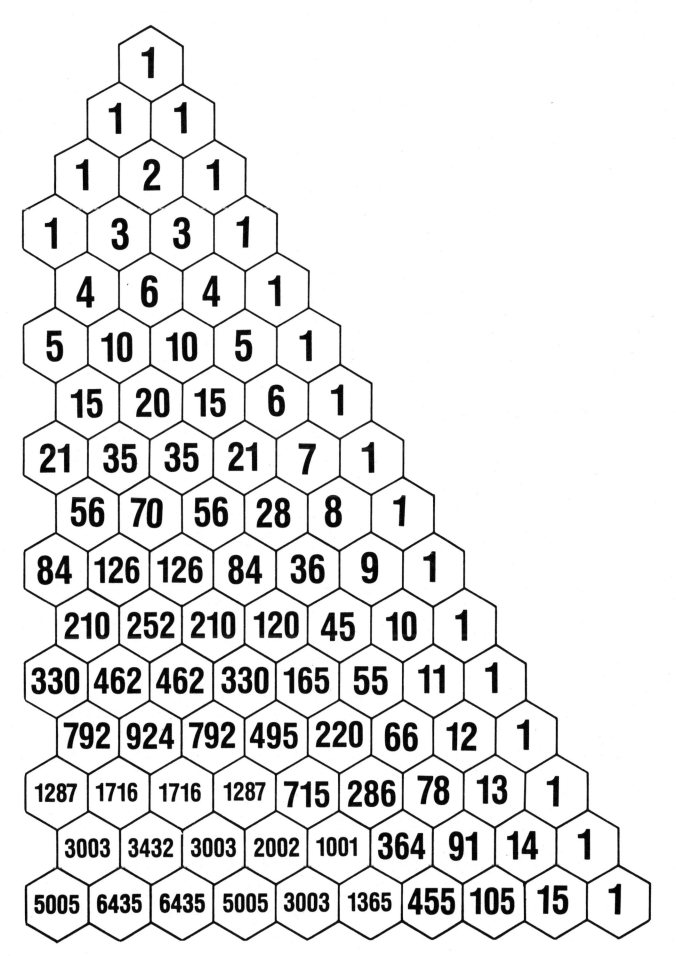

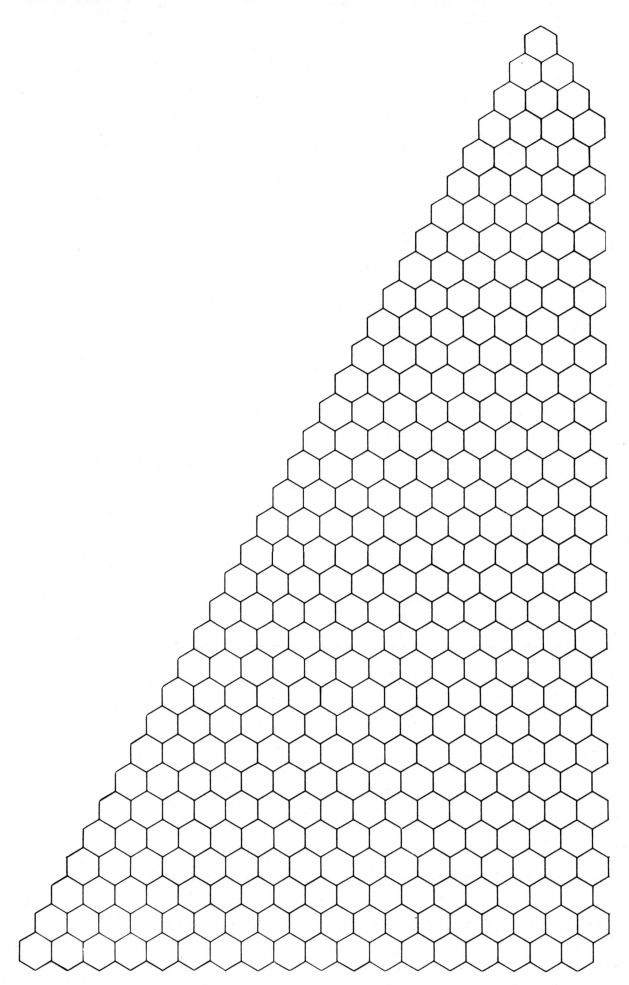

82

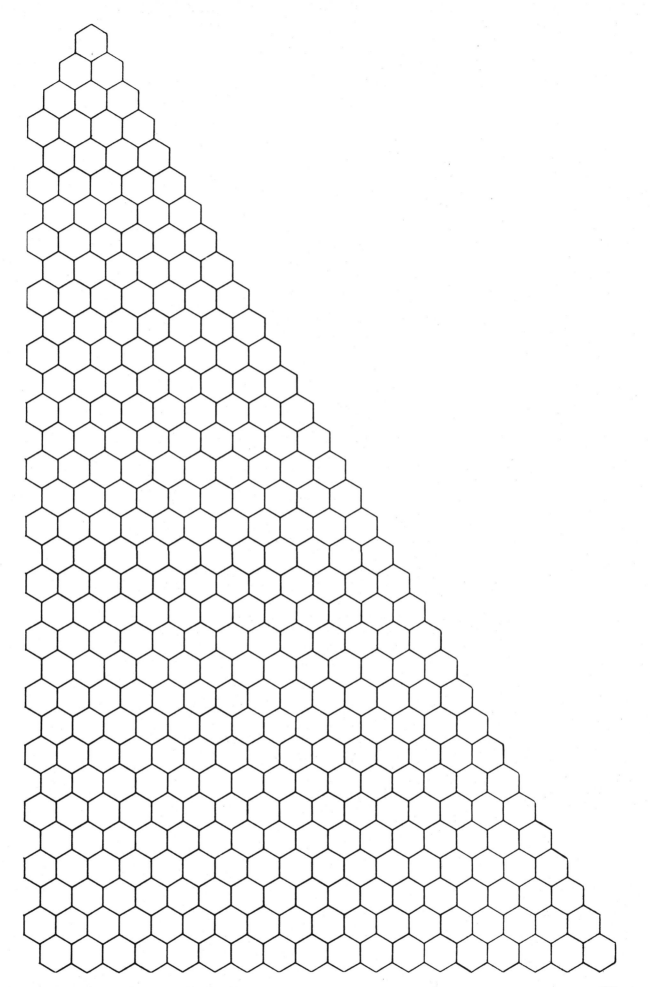

83

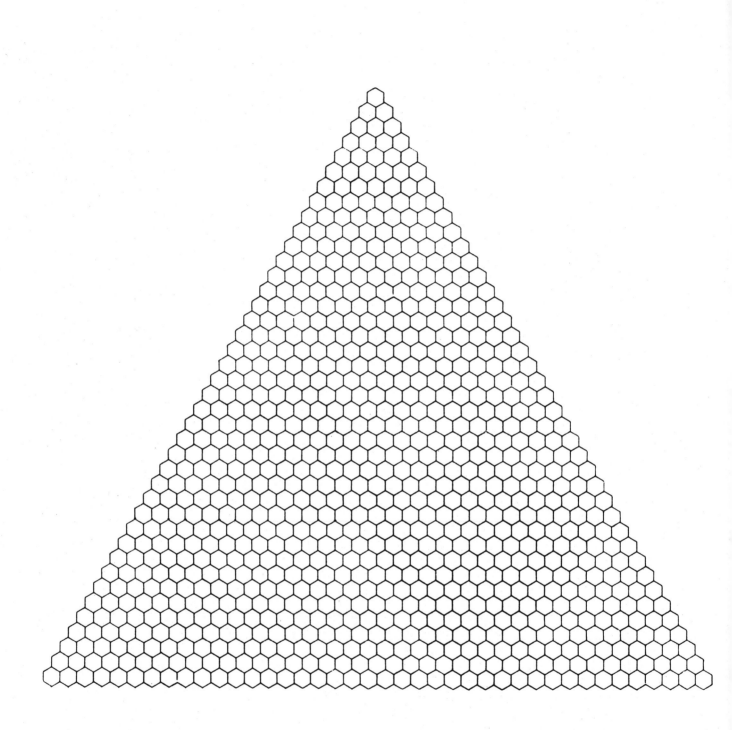

84

PRIME
FACTORIZATION CHARTS

Row 2
2 is Prime

Row 3
3 is Prime

Row 4
$4 = 2^2$
$6 = 2 \cdot 3$

Row 5
5 is Prime
$10 = 2 \cdot 5$

Row 6
$6 = 2 \cdot 3$
$15 = 3 \cdot 5$
$20 = 2^2 \cdot 5$

Row 7
7 is Prime
$21 = 3 \cdot 7$
$35 = 5 \cdot 7$

Row 8
$8 = 2^3$
$28 = 2^2 \cdot 7$
$56 = 2^3 \cdot 7$
$70 = 2 \cdot 5 \cdot 7$

Row 9
$9 = 3^2$
$36 = 2^2 \cdot 3^2$
$84 = 2^2 \cdot 3 \cdot 7$
$126 = 2 \cdot 3^2 \cdot 7$

Row 10
$10 = 2 \cdot 5$
$45 = 3^2 \cdot 5$
$120 = 2^3 \cdot 3 \cdot 5$
$210 = 2 \cdot 3 \cdot 5 \cdot 7$
$252 = 2^2 \cdot 3^2 \cdot 7$

Row 11
11 is Prime
$55 = 5 \cdot 11$
$165 = 3 \cdot 5 \cdot 11$
$330 = 2 \cdot 3 \cdot 5 \cdot 11$
$462 = 2 \cdot 3 \cdot 7 \cdot 11$

Row 12
$12 = 2^2 \cdot 3$
$66 = 2 \cdot 3 \cdot 11$
$220 = 2^2 \cdot 5 \cdot 11$
$495 = 3^2 \cdot 5 \cdot 11$
$792 = 2^3 \cdot 3^2 \cdot 11$
$924 = 2^2 \cdot 3 \cdot 7 \cdot 11$

Row 13
13 is Prime
$78 = 2 \cdot 3 \cdot 13$
$286 = 2 \cdot 11 \cdot 13$
$715 = 5 \cdot 11 \cdot 13$
$1287 = 3^2 \cdot 11 \cdot 13$
$1716 = 2^2 \cdot 3 \cdot 11 \cdot 13$

Row 14
$14 = 2 \cdot 7$
$91 = 7 \cdot 13$
$364 = 2^2 \cdot 7 \cdot 13$
$1001 = 7 \cdot 11 \cdot 13$
$2002 = 2 \cdot 7 \cdot 11 \cdot 13$
$3003 = 3 \cdot 7 \cdot 11 \cdot 13$
$3432 = 2^3 \cdot 3 \cdot 11 \cdot 13$

Row 15
$15 = 3 \cdot 5$
$105 = 3 \cdot 5 \cdot 7$
$455 = 5 \cdot 7 \cdot 13$
$1365 = 3 \cdot 5 \cdot 7 \cdot 13$
$3003 = 3 \cdot 7 \cdot 11 \cdot 13$
$5005 = 5 \cdot 7 \cdot 11 \cdot 13$
$6435 = 3^2 \cdot 5 \cdot 11 \cdot 13$

Row 16
$16 = 2^4$
$120 = 2^3 \cdot 3 \cdot 5$
$560 = 2^4 \cdot 5 \cdot 7$
$1820 = 2^2 \cdot 5 \cdot 7 \cdot 13$
$4368 = 2^4 \cdot 3 \cdot 7 \cdot 13$
$8008 = 2^3 \cdot 7 \cdot 11 \cdot 13$
$11440 = 2^4 \cdot 5 \cdot 11 \cdot 13$
$12870 = 2 \cdot 3^2 \cdot 5 \cdot 11 \cdot 13$

Row 17
17 is Prime
$136 = 2^3 \cdot 17$
$680 = 2^3 \cdot 5 \cdot 17$
$2380 = 2^2 \cdot 5 \cdot 7 \cdot 17$
$6188 = 2^2 \cdot 7 \cdot 13 \cdot 17$
$12376 = 2^3 \cdot 7 \cdot 13 \cdot 17$
$19448 = 2^3 \cdot 11 \cdot 13 \cdot 17$
$24310 = 2 \cdot 5 \cdot 11 \cdot 13 \cdot 17$

Row 18
$18 = 2 \cdot 3^2$
$153 = 3^2 \cdot 17$
$816 = 2^4 \cdot 3 \cdot 17$
$3060 = 2^2 \cdot 3^2 \cdot 5 \cdot 17$
$8568 = 2^3 \cdot 3^2 \cdot 7 \cdot 17$
$18564 = 2^2 \cdot 3 \cdot 7 \cdot 13 \cdot 17$
$31824 = 2^4 \cdot 3^2 \cdot 13 \cdot 17$
$43758 = 2 \cdot 3^2 \cdot 11 \cdot 13 \cdot 17$
$48620 = 2^2 \cdot 5 \cdot 11 \cdot 13 \cdot 17$

Row 19
19 is Prime
$171 = 3^2 \cdot 19$
$969 = 3 \cdot 17 \cdot 19$
$3\,876 = 2^2 \cdot 3 \cdot 17 \cdot 19$
$11\,628 = 2^2 \cdot 3^2 \cdot 17 \cdot 19$
$27\,132 = 2^2 \cdot 3 \cdot 7 \cdot 17 \cdot 19$
$50\,388 = 2^2 \cdot 3 \cdot 13 \cdot 17 \cdot 19$
$75\,582 = 2 \cdot 3^2 \cdot 13 \cdot 17 \cdot 19$
$92\,378 = 2 \cdot 11 \cdot 13 \cdot 17 \cdot 19$

Row 20
$20 = 2^2 \cdot 5$
$190 = 2 \cdot 5 \cdot 19$
$1\,140 = 2^2 \cdot 3 \cdot 5 \cdot 19$
$4\,845 = 3 \cdot 5 \cdot 17 \cdot 19$
$15\,504 = 2^4 \cdot 3 \cdot 17 \cdot 19$
$38\,760 = 2^3 \cdot 3 \cdot 5 \cdot 17 \cdot 19$
$77\,520 = 2^4 \cdot 3 \cdot 5 \cdot 17 \cdot 19$
$125\,970 = 2 \cdot 3 \cdot 5 \cdot 13 \cdot 17 \cdot 19$
$167\,960 = 2^3 \cdot 5 \cdot 13 \cdot 17 \cdot 19$
$184\,756 = 2^2 \cdot 11 \cdot 13 \cdot 17 \cdot 19$

Pascal's Triangle — Prime Factorization — To Center Number (omitting 1's)

Row 21

$21 = 3 \cdot 7$

$210 = 2 \cdot 3 \cdot 5 \cdot 7$

$1\,330 = 2 \cdot 5 \cdot 7 \cdot 19$

$5\,985 = 3^2 \cdot 5 \cdot 7 \cdot 19$

$20\,349 = 3^2 \cdot 7 \cdot 17 \cdot 19$

$54\,264 = 2^3 \cdot 3 \cdot 7 \cdot 17 \cdot 19$

$116\,280 = 2^3 \cdot 3^2 \cdot 5 \cdot 17 \cdot 19$

$203\,490 = 2 \cdot 3^2 \cdot 5 \cdot 7 \cdot 17 \cdot 19$

$293\,930 = 2 \cdot 5 \cdot 7 \cdot 13 \cdot 17 \cdot 19$

$352\,716 = 2^2 \cdot 3 \cdot 7 \cdot 13 \cdot 17 \cdot 19$

Row 22

$22 = 2 \cdot 11$

$231 = 3 \cdot 7 \cdot 11$

$1\,540 = 2^2 \cdot 5 \cdot 7 \cdot 11$

$7\,315 = 5 \cdot 7 \cdot 11 \cdot 19$

$26\,334 = 2 \cdot 3^2 \cdot 7 \cdot 11 \cdot 19$

$74\,613 = 3 \cdot 7 \cdot 11 \cdot 17 \cdot 19$

$170\,544 = 2^4 \cdot 3 \cdot 11 \cdot 17 \cdot 19$

$319\,770 = 2 \cdot 3^2 \cdot 5 \cdot 11 \cdot 17 \cdot 19$

$497\,420 = 2^2 \cdot 5 \cdot 7 \cdot 11 \cdot 17 \cdot 19$

$646\,646 = 2 \cdot 7 \cdot 11 \cdot 13 \cdot 17 \cdot 19$

$705\,432 = 2^3 \cdot 3 \cdot 7 \cdot 13 \cdot 17 \cdot 19$

Row 23

23 is Prime

$253 = 11 \cdot 23$

$1\,771 = 7 \cdot 11 \cdot 23$

$8\,855 = 5 \cdot 7 \cdot 11 \cdot 23$

$33\,649 = 7 \cdot 11 \cdot 19 \cdot 23$

$100\,947 = 3 \cdot 7 \cdot 11 \cdot 19 \cdot 23$

$245\,157 = 3 \cdot 11 \cdot 17 \cdot 19 \cdot 23$

$490\,314 = 2 \cdot 3 \cdot 11 \cdot 17 \cdot 19 \cdot 23$

$817\,190 = 2 \cdot 5 \cdot 11 \cdot 17 \cdot 19 \cdot 23$

$1\,144\,066 = 2 \cdot 7 \cdot 11 \cdot 17 \cdot 19 \cdot 23$

$1\,352\,078 = 2 \cdot 7 \cdot 13 \cdot 17 \cdot 19 \cdot 23$

Row 24

$24 = 2^3 \cdot 3$

$276 = 2^2 \cdot 3 \cdot 23$

$2\,024 = 2^3 \cdot 11 \cdot 23$

$10\,626 = 2 \cdot 3 \cdot 7 \cdot 11 \cdot 23$

$42\,504 = 2^3 \cdot 3 \cdot 7 \cdot 11 \cdot 23$

$134\,596 = 2^2 \cdot 7 \cdot 11 \cdot 19 \cdot 23$

$346\,104 = 2^3 \cdot 3^2 \cdot 11 \cdot 19 \cdot 23$

$735\,471 = 3^2 \cdot 11 \cdot 17 \cdot 19 \cdot 23$

$1\,307\,504 = 2^4 \cdot 11 \cdot 17 \cdot 19 \cdot 23$

$1\,961\,256 = 2^3 \cdot 3 \cdot 11 \cdot 17 \cdot 19 \cdot 23$

$2\,496\,144 = 2^4 \cdot 3 \cdot 7 \cdot 17 \cdot 19 \cdot 23$

$2\,704\,156 = 2^2 \cdot 7 \cdot 13 \cdot 17 \cdot 19 \cdot 23$

Row 25

$25 = 5^2$

$300 = 2^2 \cdot 3 \cdot 5^2$

$2\,300 = 2^2 \cdot 5^2 \cdot 23$

$12\,650 = 2 \cdot 5^2 \cdot 11 \cdot 23$

$53\,130 = 2 \cdot 3 \cdot 5 \cdot 7 \cdot 11 \cdot 23$

$177\,100 = 2^2 \cdot 5^2 \cdot 7 \cdot 11 \cdot 23$

$480\,700 = 2^2 \cdot 5^2 \cdot 11 \cdot 19 \cdot 23$

$1\,081\,575 = 3^2 \cdot 5^2 \cdot 11 \cdot 19 \cdot 23$

$2\,042\,975 = 5^2 \cdot 11 \cdot 17 \cdot 19 \cdot 23$

$3\,268\,760 = 2^3 \cdot 5 \cdot 11 \cdot 17 \cdot 19 \cdot 23$

$4\,457\,400 = 2^3 \cdot 3 \cdot 5^2 \cdot 17 \cdot 19 \cdot 23$

$5\,200\,300 = 2^2 \cdot 5^2 \cdot 7 \cdot 17 \cdot 19 \cdot 23$

Row 26

$26 = 2 \cdot 13$

$325 = 5^2 \cdot 13$

$2\,600 = 2^3 \cdot 5^2 \cdot 13$

$14\,950 = 2 \cdot 5^2 \cdot 13 \cdot 23$

$65\,780 = 2^2 \cdot 5 \cdot 11 \cdot 13 \cdot 23$

$230\,230 = 2 \cdot 5 \cdot 7 \cdot 11 \cdot 13 \cdot 23$

$657\,800 = 2^3 \cdot 5^2 \cdot 11 \cdot 13 \cdot 23$

$1\,562\,275 = 5^2 \cdot 11 \cdot 13 \cdot 19 \cdot 23$

$3\,124\,550 = 2 \cdot 5^2 \cdot 11 \cdot 13 \cdot 19 \cdot 23$

$5\,311\,735 = 5 \cdot 11 \cdot 13 \cdot 17 \cdot 19 \cdot 23$

$7\,726\,160 = 2^4 \cdot 5 \cdot 13 \cdot 17 \cdot 19 \cdot 23$

$9\,657\,700 = 2^2 \cdot 5^2 \cdot 13 \cdot 17 \cdot 19 \cdot 23$

$10\,400\,600 = 2^3 \cdot 5^2 \cdot 7 \cdot 17 \cdot 19 \cdot 23$

Row 27

$27 = 3^3$

$351 = 3^3 \cdot 13$

$2\,925 = 3^2 \cdot 5^2 \cdot 13$

$17\,550 = 2 \cdot 3^3 \cdot 5^2 \cdot 13$

$80\,730 = 2 \cdot 3^3 \cdot 5 \cdot 13 \cdot 23$

$296\,010 = 2 \cdot 3^2 \cdot 5 \cdot 11 \cdot 13 \cdot 23$

$888\,030 = 2 \cdot 3^3 \cdot 5 \cdot 11 \cdot 13 \cdot 23$

$2\,220\,075 = 3^3 \cdot 5^2 \cdot 11 \cdot 13 \cdot 23$

$4\,686\,825 = 3 \cdot 5^2 \cdot 11 \cdot 13 \cdot 19 \cdot 23$

$8\,436\,285 = 3^3 \cdot 5 \cdot 11 \cdot 13 \cdot 19 \cdot 23$

$13\,037\,895 = 3^3 \cdot 5 \cdot 13 \cdot 17 \cdot 19 \cdot 23$

$17\,383\,860 = 2^2 \cdot 3^2 \cdot 5 \cdot 13 \cdot 17 \cdot 19 \cdot 23$

$20\,058\,300 = 2^2 \cdot 3^3 \cdot 5^2 \cdot 17 \cdot 19 \cdot 23$

Row 28

$$28 = 2^2 \cdot 7$$
$$378 = 2 \cdot 3^3 \cdot 7$$
$$3\,276 = 2^2 \cdot 3^2 \cdot 7 \cdot 13$$
$$20\,475 = 3^2 \cdot 5^2 \cdot 7 \cdot 13$$
$$98\,280 = 2^3 \cdot 3^3 \cdot 5 \cdot 7 \cdot 13$$
$$376\,740 = 2^2 \cdot 3^2 \cdot 5 \cdot 7 \cdot 13 \cdot 23$$
$$1\,184\,040 = 2^3 \cdot 3^2 \cdot 5 \cdot 11 \cdot 13 \cdot 23$$
$$3\,108\,105 = 3^3 \cdot 5 \cdot 7 \cdot 11 \cdot 13 \cdot 23$$
$$6\,906\,900 = 2^2 \cdot 3 \cdot 5^2 \cdot 7 \cdot 11 \cdot 13 \cdot 23$$
$$13\,123\,110 = 2 \cdot 3 \cdot 5 \cdot 7 \cdot 11 \cdot 13 \cdot 19 \cdot 23$$
$$21\,474\,180 = 2^2 \cdot 3^3 \cdot 5 \cdot 7 \cdot 13 \cdot 19 \cdot 23$$
$$30\,421\,755 = 3^2 \cdot 5 \cdot 7 \cdot 13 \cdot 17 \cdot 19 \cdot 23$$
$$37\,442\,160 = 2^4 \cdot 3^2 \cdot 5 \cdot 7 \cdot 17 \cdot 19 \cdot 23$$
$$40\,116\,600 = 2^3 \cdot 3^3 \cdot 5^2 \cdot 17 \cdot 19 \cdot 23$$

Row 29

$$29 \text{ is Prime}$$
$$406 = 2 \cdot 7 \cdot 29$$
$$3\,654 = 2 \cdot 3^2 \cdot 7 \cdot 29$$
$$23\,751 = 3^2 \cdot 7 \cdot 13 \cdot 29$$
$$118\,755 = 3^2 \cdot 5 \cdot 7 \cdot 13 \cdot 29$$
$$475\,020 = 2^2 \cdot 3^2 \cdot 5 \cdot 7 \cdot 13 \cdot 29$$
$$1\,560\,780 = 2^2 \cdot 3^2 \cdot 5 \cdot 13 \cdot 23 \cdot 29$$
$$4\,292\,145 = 3^2 \cdot 5 \cdot 11 \cdot 13 \cdot 23 \cdot 29$$
$$10\,015\,005 = 3 \cdot 5 \cdot 7 \cdot 11 \cdot 13 \cdot 23 \cdot 29$$
$$20\,030\,010 = 2 \cdot 3 \cdot 5 \cdot 7 \cdot 11 \cdot 13 \cdot 23 \cdot 29$$
$$34\,597\,290 = 2 \cdot 3 \cdot 5 \cdot 7 \cdot 13 \cdot 19 \cdot 23 \cdot 29$$
$$51\,895\,935 = 3^2 \cdot 5 \cdot 7 \cdot 13 \cdot 19 \cdot 23 \cdot 29$$
$$67\,863\,915 = 3^2 \cdot 5 \cdot 7 \cdot 17 \cdot 19 \cdot 23 \cdot 29$$
$$77\,558\,760 = 2^3 \cdot 3^2 \cdot 5 \cdot 17 \cdot 19 \cdot 23 \cdot 29$$

Row 30

$$30 = 2 \cdot 3 \cdot 5$$
$$435 = 3 \cdot 5 \cdot 29$$
$$4\,060 = 2^2 \cdot 5 \cdot 7 \cdot 29$$
$$27\,405 = 3^3 \cdot 5 \cdot 7 \cdot 29$$
$$142\,506 = 2 \cdot 3^3 \cdot 7 \cdot 13 \cdot 29$$
$$593\,775 = 3^2 \cdot 5^2 \cdot 7 \cdot 13 \cdot 29$$
$$2\,035\,800 = 2^3 \cdot 3^3 \cdot 5^2 \cdot 13 \cdot 29$$
$$5\,852\,925 = 3^3 \cdot 5^2 \cdot 13 \cdot 23 \cdot 29$$
$$14\,307\,150 = 2 \cdot 3 \cdot 5^2 \cdot 11 \cdot 13 \cdot 23 \cdot 29$$
$$30\,045\,015 = 3^2 \cdot 5 \cdot 7 \cdot 11 \cdot 13 \cdot 23 \cdot 29$$
$$54\,627\,300 = 2^2 \cdot 3^2 \cdot 5^2 \cdot 7 \cdot 13 \cdot 23 \cdot 29$$
$$86\,493\,225 = 3 \cdot 5^2 \cdot 7 \cdot 13 \cdot 19 \cdot 23 \cdot 29$$
$$119\,759\,850 = 2 \cdot 3^3 \cdot 5^2 \cdot 7 \cdot 19 \cdot 23 \cdot 29$$
$$145\,422\,675 = 3^3 \cdot 5^2 \cdot 17 \cdot 19 \cdot 23 \cdot 29$$
$$155\,117\,520 = 2^4 \cdot 3^2 \cdot 5 \cdot 17 \cdot 19 \cdot 23 \cdot 29$$

Row 31

$$31 \text{ is Prime}$$
$$465 = 3 \cdot 5 \cdot 31$$
$$4\,495 = 5 \cdot 29 \cdot 31$$
$$31\,465 = 5 \cdot 7 \cdot 29 \cdot 31$$
$$169\,911 = 3^3 \cdot 7 \cdot 29 \cdot 31$$
$$736\,281 = 3^2 \cdot 7 \cdot 13 \cdot 29 \cdot 31$$
$$2\,629\,575 = 3^2 \cdot 5^2 \cdot 13 \cdot 29 \cdot 31$$
$$7\,888\,725 = 3^3 \cdot 5^2 \cdot 13 \cdot 29 \cdot 31$$
$$20\,160\,075 = 3 \cdot 5^2 \cdot 13 \cdot 23 \cdot 29 \cdot 31$$
$$44\,352\,165 = 3 \cdot 5 \cdot 11 \cdot 13 \cdot 23 \cdot 29 \cdot 31$$
$$84\,672\,315 = 3^2 \cdot 5 \cdot 7 \cdot 13 \cdot 23 \cdot 29 \cdot 31$$
$$141\,120\,525 = 3 \cdot 5^2 \cdot 7 \cdot 13 \cdot 23 \cdot 29 \cdot 31$$
$$206\,253\,075 = 3 \cdot 5^2 \cdot 7 \cdot 19 \cdot 23 \cdot 29 \cdot 31$$
$$265\,182\,525 = 3^3 \cdot 5^2 \cdot 19 \cdot 23 \cdot 29 \cdot 31$$
$$300\,540\,195 = 3^2 \cdot 5 \cdot 17 \cdot 19 \cdot 23 \cdot 29 \cdot 31$$

Row 32

$$32 = 2^5$$
$$496 = 2^4 \cdot 31$$
$$4\,960 = 2^5 \cdot 5 \cdot 31$$
$$35\,960 = 2^3 \cdot 5 \cdot 29 \cdot 31$$
$$201\,376 = 2^5 \cdot 7 \cdot 29 \cdot 31$$
$$906\,192 = 2^4 \cdot 3^2 \cdot 7 \cdot 29 \cdot 31$$
$$3\,365\,856 = 2^5 \cdot 3^2 \cdot 13 \cdot 29 \cdot 31$$
$$10\,518\,300 = 2^2 \cdot 3^2 \cdot 5^2 \cdot 13 \cdot 29 \cdot 31$$
$$28\,048\,800 = 2^5 \cdot 3 \cdot 5^2 \cdot 13 \cdot 29 \cdot 31$$
$$64\,512\,240 = 2^4 \cdot 3 \cdot 5 \cdot 13 \cdot 23 \cdot 29 \cdot 31$$
$$129\,024\,480 = 2^5 \cdot 3 \cdot 5 \cdot 13 \cdot 23 \cdot 29 \cdot 31$$
$$225\,792\,840 = 2^3 \cdot 3 \cdot 5 \cdot 7 \cdot 13 \cdot 23 \cdot 29 \cdot 31$$
$$347\,373\,600 = 2^5 \cdot 3 \cdot 5^2 \cdot 7 \cdot 23 \cdot 29 \cdot 31$$
$$471\,435\,600 = 2^4 \cdot 3 \cdot 5^2 \cdot 19 \cdot 23 \cdot 29 \cdot 31$$
$$565\,722\,720 = 2^5 \cdot 3^2 \cdot 5 \cdot 19 \cdot 23 \cdot 29 \cdot 31$$
$$601\,080\,390 = 2 \cdot 3^2 \cdot 5 \cdot 17 \cdot 19 \cdot 23 \cdot 29 \cdot 31$$

Row 33

$$33 = 3 \cdot 11$$
$$528 = 2^4 \cdot 3 \cdot 11$$
$$5\,456 = 2^4 \cdot 11 \cdot 31$$
$$40\,920 = 2^3 \cdot 3 \cdot 5 \cdot 11 \cdot 31$$
$$237\,336 = 2^3 \cdot 3 \cdot 11 \cdot 29 \cdot 31$$
$$1\,107\,568 = 2^4 \cdot 7 \cdot 11 \cdot 29 \cdot 31$$
$$4\,272\,048 = 2^4 \cdot 3^3 \cdot 11 \cdot 29 \cdot 31$$
$$13\,884\,156 = 2^2 \cdot 3^3 \cdot 11 \cdot 13 \cdot 29 \cdot 31$$
$$38\,567\,100 = 2^2 \cdot 3 \cdot 5^2 \cdot 11 \cdot 13 \cdot 29 \cdot 31$$
$$92\,561\,040 = 2^4 \cdot 3^2 \cdot 5 \cdot 11 \cdot 13 \cdot 29 \cdot 31$$
$$193\,536\,720 = 2^4 \cdot 3^2 \cdot 5 \cdot 13 \cdot 23 \cdot 29 \cdot 31$$
$$354\,817\,320 = 2^3 \cdot 3 \cdot 5 \cdot 11 \cdot 13 \cdot 23 \cdot 29 \cdot 31$$
$$573\,166\,440 = 2^3 \cdot 3^2 \cdot 5 \cdot 7 \cdot 11 \cdot 23 \cdot 29 \cdot 31$$
$$818\,809\,200 = 2^4 \cdot 3^2 \cdot 5^2 \cdot 11 \cdot 23 \cdot 29 \cdot 31$$
$$1\,037\,158\,320 = 2^4 \cdot 3 \cdot 5 \cdot 11 \cdot 19 \cdot 23 \cdot 29 \cdot 31$$
$$1\,166\,803\,110 = 2 \cdot 3^3 \cdot 5 \cdot 11 \cdot 19 \cdot 23 \cdot 29 \cdot 31$$

Row 34

$$34 = 2 \cdot 17$$
$$561 = 3 \cdot 11 \cdot 17$$
$$5\,984 = 2^5 \cdot 11 \cdot 17$$
$$46\,376 = 2^3 \cdot 11 \cdot 17 \cdot 31$$
$$278\,256 = 2^4 \cdot 3 \cdot 11 \cdot 17 \cdot 31$$
$$1\,344\,904 = 2^3 \cdot 11 \cdot 17 \cdot 29 \cdot 31$$
$$6\,370\,616 = 2^5 \cdot 11 \cdot 17 \cdot 29 \cdot 31$$
$$18\,156\,204 = 2^2 \cdot 3^3 \cdot 11 \cdot 17 \cdot 29 \cdot 31$$
$$52\,451\,256 = 2^3 \cdot 3 \cdot 11 \cdot 13 \cdot 17 \cdot 29 \cdot 31$$
$$131\,128\,140 = 2^2 \cdot 3 \cdot 5 \cdot 11 \cdot 13 \cdot 17 \cdot 29 \cdot 31$$
$$286\,097\,760 = 2^5 \cdot 3^2 \cdot 5 \cdot 13 \cdot 17 \cdot 29 \cdot 31$$
$$548\,354\,040 = 2^3 \cdot 3 \cdot 5 \cdot 13 \cdot 17 \cdot 23 \cdot 29 \cdot 31$$
$$927\,983\,760 = 2^4 \cdot 3 \cdot 5 \cdot 11 \cdot 17 \cdot 23 \cdot 29 \cdot 31$$
$$1\,391\,975\,640 = 2^3 \cdot 3^2 \cdot 5 \cdot 11 \cdot 17 \cdot 23 \cdot 29 \cdot 31$$
$$1\,855\,967\,520 = 2^5 \cdot 3 \cdot 5 \cdot 11 \cdot 17 \cdot 23 \cdot 29 \cdot 31$$
$$2\,203\,961\,430 = 2 \cdot 3 \cdot 5 \cdot 11 \cdot 17 \cdot 19 \cdot 23 \cdot 29 \cdot 31$$
$$2\,333\,606\,220 = 2^2 \cdot 3^3 \cdot 5 \cdot 11 \cdot 19 \cdot 23 \cdot 29 \cdot 31$$

Row 35

$$35 = 5 \cdot 7$$
$$595 = 5 \cdot 7 \cdot 17$$
$$6\,545 = 5 \cdot 7 \cdot 11 \cdot 17$$
$$52\,360 = 2^3 \cdot 5 \cdot 7 \cdot 11 \cdot 17$$
$$324\,632 = 2^3 \cdot 7 \cdot 11 \cdot 17 \cdot 31$$
$$1\,623\,160 = 2^3 \cdot 5 \cdot 7 \cdot 11 \cdot 17 \cdot 31$$
$$6\,724\,520 = 2^3 \cdot 5 \cdot 11 \cdot 17 \cdot 29 \cdot 31$$
$$23\,535\,820 = 2^2 \cdot 5 \cdot 7 \cdot 11 \cdot 17 \cdot 29 \cdot 31$$
$$70\,607\,460 = 2^2 \cdot 3 \cdot 5 \cdot 7 \cdot 11 \cdot 17 \cdot 29 \cdot 31$$
$$183\,579\,396 = 2^2 \cdot 3 \cdot 7 \cdot 11 \cdot 13 \cdot 17 \cdot 29 \cdot 31$$
$$417\,225\,900 = 2^2 \cdot 3 \cdot 5^2 \cdot 7 \cdot 13 \cdot 17 \cdot 29 \cdot 31$$
$$834\,451\,800 = 2^3 \cdot 3 \cdot 5^2 \cdot 7 \cdot 13 \cdot 17 \cdot 29 \cdot 31$$
$$1\,476\,337\,800 = 2^3 \cdot 3 \cdot 5^2 \cdot 7 \cdot 17 \cdot 23 \cdot 29 \cdot 31$$
$$2\,319\,959\,400 = 2^3 \cdot 3 \cdot 5^2 \cdot 11 \cdot 17 \cdot 23 \cdot 29 \cdot 31$$
$$3\,247\,943\,160 = 2^3 \cdot 3 \cdot 5 \cdot 7 \cdot 11 \cdot 17 \cdot 23 \cdot 29 \cdot 31$$
$$4\,059\,928\,950 = 2 \cdot 3 \cdot 5^2 \cdot 7 \cdot 11 \cdot 17 \cdot 23 \cdot 29 \cdot 31$$
$$4\,537\,567\,650 = 2 \cdot 3 \cdot 5^2 \cdot 7 \cdot 11 \cdot 19 \cdot 23 \cdot 29 \cdot 31$$

Row 36

$$36 = 2^2 \cdot 3^2$$
$$630 = 2 \cdot 3^2 \cdot 5 \cdot 7$$
$$7\,140 = 2^2 \cdot 3 \cdot 5 \cdot 7 \cdot 17$$
$$58\,905 = 3^2 \cdot 5 \cdot 7 \cdot 11 \cdot 17$$
$$376\,992 = 2^5 \cdot 3^2 \cdot 7 \cdot 11 \cdot 17$$
$$1\,947\,792 = 2^4 \cdot 3 \cdot 7 \cdot 11 \cdot 17 \cdot 31$$
$$8\,347\,680 = 2^5 \cdot 3^2 \cdot 5 \cdot 11 \cdot 17 \cdot 31$$
$$30\,260\,340 = 2^2 \cdot 3^2 \cdot 5 \cdot 11 \cdot 17 \cdot 29 \cdot 31$$
$$94\,143\,280 = 2^4 \cdot 5 \cdot 7 \cdot 11 \cdot 17 \cdot 29 \cdot 31$$
$$254\,186\,856 = 2^3 \cdot 3^3 \cdot 7 \cdot 11 \cdot 17 \cdot 29 \cdot 31$$
$$600\,805\,296 = 2^4 \cdot 3^3 \cdot 7 \cdot 13 \cdot 17 \cdot 29 \cdot 31$$
$$1\,251\,677\,700 = 2^2 \cdot 3^2 \cdot 5^2 \cdot 7 \cdot 13 \cdot 17 \cdot 29 \cdot 31$$
$$2\,310\,789\,600 = 2^5 \cdot 3^3 \cdot 5^2 \cdot 7 \cdot 17 \cdot 29 \cdot 31$$
$$3\,796\,297\,200 = 2^4 \cdot 3^3 \cdot 5^2 \cdot 17 \cdot 23 \cdot 29 \cdot 31$$
$$5\,567\,902\,560 = 2^5 \cdot 3^2 \cdot 5 \cdot 11 \cdot 17 \cdot 23 \cdot 29 \cdot 31$$
$$7\,307\,872\,110 = 2 \cdot 3^3 \cdot 5 \cdot 7 \cdot 11 \cdot 17 \cdot 23 \cdot 29 \cdot 31$$
$$8\,597\,496\,600 = 2^3 \cdot 3^3 \cdot 5^2 \cdot 7 \cdot 11 \cdot 23 \cdot 29 \cdot 31$$
$$9\,075\,135\,300 = 2^2 \cdot 3 \cdot 5^2 \cdot 7 \cdot 11 \cdot 19 \cdot 23 \cdot 29 \cdot 31$$

Row 37

$$37 \text{ is Prime}$$
$$666 = 2 \cdot 3^2 \cdot 37$$
$$7\,770 = 2 \cdot 3 \cdot 5 \cdot 7 \cdot 37$$
$$66\,045 = 3 \cdot 5 \cdot 7 \cdot 17 \cdot 37$$
$$435\,897 = 3^2 \cdot 7 \cdot 11 \cdot 17 \cdot 37$$
$$2\,324\,784 = 2^4 \cdot 3 \cdot 7 \cdot 11 \cdot 17 \cdot 37$$
$$10\,295\,472 = 2^4 \cdot 3 \cdot 11 \cdot 17 \cdot 31 \cdot 37$$
$$38\,608\,020 = 2^2 \cdot 3^2 \cdot 5 \cdot 11 \cdot 17 \cdot 31 \cdot 37$$
$$124\,403\,620 = 2^2 \cdot 5 \cdot 11 \cdot 17 \cdot 29 \cdot 31 \cdot 37$$
$$348\,330\,136 = 2^3 \cdot 7 \cdot 11 \cdot 17 \cdot 29 \cdot 31 \cdot 37$$
$$854\,992\,152 = 2^3 \cdot 3^3 \cdot 7 \cdot 17 \cdot 29 \cdot 31 \cdot 37$$
$$1\,852\,482\,996 = 2^2 \cdot 3^2 \cdot 7 \cdot 13 \cdot 17 \cdot 29 \cdot 31 \cdot 37$$
$$3\,562\,467\,300 = 2^2 \cdot 3^2 \cdot 5^2 \cdot 7 \cdot 17 \cdot 29 \cdot 31 \cdot 37$$
$$6\,107\,086\,800 = 2^4 \cdot 3^3 \cdot 5^2 \cdot 17 \cdot 29 \cdot 31 \cdot 37$$
$$9\,364\,199\,760 = 2^4 \cdot 3^2 \cdot 5 \cdot 17 \cdot 23 \cdot 29 \cdot 31 \cdot 37$$
$$12\,875\,774\,670 = 2 \cdot 3^2 \cdot 5 \cdot 11 \cdot 17 \cdot 23 \cdot 29 \cdot 31 \cdot 37$$
$$15\,905\,368\,710 = 2 \cdot 3^3 \cdot 5 \cdot 7 \cdot 11 \cdot 23 \cdot 29 \cdot 31 \cdot 37$$
$$17\,672\,631\,900 = 2^2 \cdot 3 \cdot 5^2 \cdot 7 \cdot 11 \cdot 23 \cdot 29 \cdot 31 \cdot 37$$

Row 38

$$
\begin{aligned}
38 &= 2 \cdot 19 \\
703 &= 19 \cdot 37 \\
8\,436 &= 2^2 \cdot 3 \cdot 19 \cdot 37 \\
73\,815 &= 3 \cdot 5 \cdot 7 \cdot 19 \cdot 37 \\
501\,942 &= 2 \cdot 3 \cdot 7 \cdot 17 \cdot 19 \cdot 37 \\
2\,760\,681 &= 3 \cdot 7 \cdot 11 \cdot 17 \cdot 19 \cdot 37 \\
12\,620\,256 &= 2^5 \cdot 3 \cdot 11 \cdot 17 \cdot 19 \cdot 37 \\
48\,903\,492 &= 2^2 \cdot 3 \cdot 11 \cdot 17 \cdot 19 \cdot 31 \cdot 37 \\
163\,011\,640 &= 2^3 \cdot 5 \cdot 11 \cdot 17 \cdot 19 \cdot 31 \cdot 37 \\
472\,733\,756 &= 2^2 \cdot 11 \cdot 17 \cdot 19 \cdot 29 \cdot 31 \cdot 37 \\
1\,203\,322\,288 &= 2^4 \cdot 7 \cdot 17 \cdot 19 \cdot 29 \cdot 31 \cdot 37 \\
2\,707\,475\,148 &= 2^2 \cdot 3^2 \cdot 7 \cdot 17 \cdot 19 \cdot 29 \cdot 31 \cdot 37 \\
5\,414\,950\,296 &= 2^3 \cdot 3^2 \cdot 7 \cdot 17 \cdot 19 \cdot 29 \cdot 31 \cdot 37 \\
9\,669\,554\,100 &= 2^2 \cdot 3^2 \cdot 5^2 \cdot 17 \cdot 19 \cdot 29 \cdot 31 \cdot 37 \\
15\,471\,286\,560 &= 2^5 \cdot 3^2 \cdot 5 \cdot 17 \cdot 19 \cdot 29 \cdot 31 \cdot 37 \\
22\,239\,974\,430 &= 2 \cdot 3^2 \cdot 5 \cdot 17 \cdot 19 \cdot 23 \cdot 29 \cdot 31 \cdot 37 \\
28\,781\,143\,380 &= 2^2 \cdot 3^2 \cdot 5 \cdot 11 \cdot 19 \cdot 23 \cdot 29 \cdot 31 \cdot 37 \\
33\,578\,000\,610 &= 2 \cdot 3 \cdot 5 \cdot 7 \cdot 11 \cdot 19 \cdot 23 \cdot 29 \cdot 31 \cdot 37 \\
35\,345\,263\,800 &= 2^3 \cdot 3 \cdot 5^2 \cdot 7 \cdot 11 \cdot 23 \cdot 29 \cdot 31 \cdot 37
\end{aligned}
$$

Row 39

$$
\begin{aligned}
39 &= 3 \cdot 13 \\
741 &= 3 \cdot 13 \cdot 19 \\
9\,139 &= 13 \cdot 19 \cdot 37 \\
82\,251 &= 3^2 \cdot 13 \cdot 19 \cdot 37 \\
575\,757 &= 3^2 \cdot 7 \cdot 13 \cdot 19 \cdot 37 \\
3\,262\,623 &= 3 \cdot 7 \cdot 13 \cdot 17 \cdot 19 \cdot 37 \\
15\,380\,937 &= 3^2 \cdot 11 \cdot 13 \cdot 17 \cdot 19 \cdot 37 \\
61\,523\,748 &= 2^2 \cdot 3^2 \cdot 11 \cdot 13 \cdot 17 \cdot 19 \cdot 37 \\
211\,915\,132 &= 2^2 \cdot 11 \cdot 13 \cdot 17 \cdot 19 \cdot 31 \cdot 37 \\
635\,745\,396 &= 2^2 \cdot 3 \cdot 11 \cdot 13 \cdot 17 \cdot 19 \cdot 31 \cdot 37 \\
1\,676\,056\,044 &= 2^2 \cdot 3 \cdot 13 \cdot 17 \cdot 19 \cdot 29 \cdot 31 \cdot 37 \\
3\,910\,797\,436 &= 2^2 \cdot 7 \cdot 13 \cdot 17 \cdot 19 \cdot 29 \cdot 31 \cdot 37 \\
8\,122\,425\,444 &= 2^2 \cdot 3^3 \cdot 7 \cdot 17 \cdot 19 \cdot 29 \cdot 31 \cdot 37 \\
15\,084\,504\,396 &= 2^2 \cdot 3^3 \cdot 13 \cdot 17 \cdot 19 \cdot 29 \cdot 31 \cdot 37 \\
25\,140\,840\,660 &= 2^2 \cdot 3^2 \cdot 5 \cdot 13 \cdot 17 \cdot 19 \cdot 29 \cdot 31 \cdot 37 \\
37\,711\,260\,990 &= 2 \cdot 3^3 \cdot 5 \cdot 13 \cdot 17 \cdot 19 \cdot 29 \cdot 31 \cdot 37 \\
51\,021\,117\,810 &= 2 \cdot 3^3 \cdot 5 \cdot 13 \cdot 19 \cdot 23 \cdot 29 \cdot 31 \cdot 37 \\
62\,359\,143\,990 &= 2 \cdot 3 \cdot 5 \cdot 11 \cdot 13 \cdot 19 \cdot 23 \cdot 29 \cdot 31 \cdot 37 \\
68\,923\,264\,410 &= 2 \cdot 3^2 \cdot 5 \cdot 7 \cdot 11 \cdot 13 \cdot 23 \cdot 29 \cdot 31 \cdot 37
\end{aligned}
$$

Row 40

$$40 = 2^3 \cdot 5$$
$$780 = 2^2 \cdot 3 \cdot 5 \cdot 13$$
$$9\,880 = 2^3 \cdot 5 \cdot 13 \cdot 19$$
$$91\,390 = 2 \cdot 5 \cdot 13 \cdot 19 \cdot 37$$
$$658\,008 = 2^3 \cdot 3^2 \cdot 13 \cdot 19 \cdot 37$$
$$3\,838\,380 = 2^2 \cdot 3 \cdot 5 \cdot 7 \cdot 13 \cdot 19 \cdot 37$$
$$18\,643\,560 = 2^3 \cdot 3 \cdot 5 \cdot 13 \cdot 17 \cdot 19 \cdot 37$$
$$76\,904\,685 = 3^2 \cdot 5 \cdot 11 \cdot 13 \cdot 17 \cdot 19 \cdot 37$$
$$273\,438\,880 = 2^5 \cdot 5 \cdot 11 \cdot 13 \cdot 17 \cdot 19 \cdot 37$$
$$847\,660\,528 = 2^4 \cdot 11 \cdot 13 \cdot 17 \cdot 19 \cdot 31 \cdot 37$$
$$2\,311\,801\,440 = 2^5 \cdot 3 \cdot 5 \cdot 13 \cdot 17 \cdot 19 \cdot 31 \cdot 37$$
$$5\,586\,853\,480 = 2^3 \cdot 5 \cdot 13 \cdot 17 \cdot 19 \cdot 29 \cdot 31 \cdot 37$$
$$12\,033\,222\,880 = 2^5 \cdot 5 \cdot 7 \cdot 17 \cdot 19 \cdot 29 \cdot 31 \cdot 37$$
$$23\,206\,929\,840 = 2^4 \cdot 3^3 \cdot 5 \cdot 17 \cdot 19 \cdot 29 \cdot 31 \cdot 37$$
$$40\,225\,345\,056 = 2^5 \cdot 3^2 \cdot 13 \cdot 17 \cdot 19 \cdot 29 \cdot 31 \cdot 37$$
$$62\,852\,101\,650 = 2 \cdot 3^2 \cdot 5^2 \cdot 13 \cdot 17 \cdot 19 \cdot 29 \cdot 31 \cdot 37$$
$$88\,732\,378\,800 = 2^4 \cdot 3^3 \cdot 5^2 \cdot 13 \cdot 19 \cdot 29 \cdot 31 \cdot 37$$
$$113\,380\,261\,800 = 2^3 \cdot 3 \cdot 5^2 \cdot 13 \cdot 19 \cdot 23 \cdot 29 \cdot 31 \cdot 37$$
$$131\,282\,408\,400 = 2^4 \cdot 3 \cdot 5^2 \cdot 11 \cdot 13 \cdot 23 \cdot 29 \cdot 31 \cdot 37$$
$$137\,846\,528\,820 = 2^2 \cdot 3^2 \cdot 5 \cdot 7 \cdot 11 \cdot 13 \cdot 23 \cdot 29 \cdot 31 \cdot 37$$

Row 41

41 is Prime
$$820 = 2^2 \cdot 5 \cdot 41$$
$$10\,660 = 2^2 \cdot 5 \cdot 13 \cdot 41$$
$$101\,270 = 2 \cdot 5 \cdot 13 \cdot 19 \cdot 41$$
$$749\,398 = 2 \cdot 13 \cdot 19 \cdot 37 \cdot 41$$
$$4\,496\,388 = 2^2 \cdot 3 \cdot 13 \cdot 19 \cdot 37 \cdot 41$$
$$22\,481\,940 = 2^2 \cdot 3 \cdot 5 \cdot 13 \cdot 19 \cdot 37 \cdot 41$$
$$95\,548\,245 = 3 \cdot 5 \cdot 13 \cdot 17 \cdot 19 \cdot 37 \cdot 41$$
$$350\,343\,565 = 5 \cdot 11 \cdot 13 \cdot 17 \cdot 19 \cdot 37 \cdot 41$$
$$1\,121\,099\,408 = 2^4 \cdot 11 \cdot 13 \cdot 17 \cdot 19 \cdot 37 \cdot 41$$
$$3\,159\,461\,968 = 2^4 \cdot 13 \cdot 17 \cdot 19 \cdot 31 \cdot 37 \cdot 41$$
$$7\,898\,654\,920 = 2^3 \cdot 5 \cdot 13 \cdot 17 \cdot 19 \cdot 31 \cdot 37 \cdot 41$$
$$17\,620\,076\,360 = 2^3 \cdot 5 \cdot 17 \cdot 19 \cdot 29 \cdot 31 \cdot 37 \cdot 41$$
$$35\,240\,152\,720 = 2^4 \cdot 5 \cdot 17 \cdot 19 \cdot 29 \cdot 31 \cdot 37 \cdot 41$$
$$63\,432\,274\,896 = 2^4 \cdot 3^2 \cdot 17 \cdot 19 \cdot 29 \cdot 31 \cdot 37 \cdot 41$$
$$103\,077\,446\,706 = 2 \cdot 3^2 \cdot 13 \cdot 17 \cdot 19 \cdot 29 \cdot 31 \cdot 37 \cdot 41$$
$$151\,584\,480\,450 = 2 \cdot 3^2 \cdot 5^2 \cdot 13 \cdot 19 \cdot 29 \cdot 31 \cdot 37 \cdot 41$$
$$202\,112\,640\,600 = 2^3 \cdot 3 \cdot 5^2 \cdot 13 \cdot 19 \cdot 29 \cdot 31 \cdot 37 \cdot 41$$
$$244\,662\,670\,200 = 2^3 \cdot 3 \cdot 5^2 \cdot 13 \cdot 23 \cdot 29 \cdot 31 \cdot 37 \cdot 41$$
$$269\,128\,937\,220 = 2^2 \cdot 3 \cdot 5 \cdot 11 \cdot 13 \cdot 23 \cdot 29 \cdot 31 \cdot 37 \cdot 41$$

Row 42

$$42 = 2 \cdot 3 \cdot 7$$
$$861 = 3 \cdot 7 \cdot 41$$
$$11\,480 = 2^3 \cdot 5 \cdot 7 \cdot 41$$
$$111\,930 = 2 \cdot 3 \cdot 5 \cdot 7 \cdot 13 \cdot 41$$
$$850\,668 = 2^2 \cdot 3 \cdot 7 \cdot 13 \cdot 19 \cdot 41$$
$$5\,245\,786 = 2 \cdot 7 \cdot 13 \cdot 19 \cdot 37 \cdot 41$$
$$26\,978\,328 = 2^3 \cdot 3^2 \cdot 13 \cdot 19 \cdot 37 \cdot 41$$
$$118\,030\,185 = 3^2 \cdot 5 \cdot 7 \cdot 13 \cdot 19 \cdot 37 \cdot 41$$
$$445\,891\,810 = 2 \cdot 5 \cdot 7 \cdot 13 \cdot 17 \cdot 19 \cdot 37 \cdot 41$$
$$1\,471\,442\,973 = 3 \cdot 7 \cdot 11 \cdot 13 \cdot 17 \cdot 19 \cdot 37 \cdot 41$$
$$4\,280\,561\,376 = 2^5 \cdot 3 \cdot 7 \cdot 13 \cdot 17 \cdot 19 \cdot 37 \cdot 41$$
$$11\,058\,116\,888 = 2^3 \cdot 7 \cdot 13 \cdot 17 \cdot 19 \cdot 31 \cdot 37 \cdot 41$$
$$25\,518\,731\,280 = 2^4 \cdot 3 \cdot 5 \cdot 7 \cdot 17 \cdot 19 \cdot 31 \cdot 37 \cdot 41$$
$$52\,860\,229\,080 = 2^3 \cdot 3 \cdot 5 \cdot 17 \cdot 19 \cdot 29 \cdot 31 \cdot 37 \cdot 41$$
$$98\,672\,427\,616 = 2^5 \cdot 7 \cdot 17 \cdot 19 \cdot 29 \cdot 31 \cdot 37 \cdot 41$$
$$166\,509\,721\,602 = 2 \cdot 3^3 \cdot 7 \cdot 17 \cdot 19 \cdot 29 \cdot 31 \cdot 37 \cdot 41$$
$$254\,661\,927\,156 = 2^2 \cdot 3^3 \cdot 7 \cdot 13 \cdot 19 \cdot 29 \cdot 31 \cdot 37 \cdot 41$$
$$353\,697\,121\,050 = 2 \cdot 3 \cdot 5^2 \cdot 7 \cdot 13 \cdot 19 \cdot 29 \cdot 31 \cdot 37 \cdot 41$$
$$446\,775\,310\,800 = 2^4 \cdot 3^2 \cdot 5^2 \cdot 7 \cdot 13 \cdot 29 \cdot 31 \cdot 37 \cdot 41$$
$$513\,791\,607\,420 = 2^2 \cdot 3^2 \cdot 5 \cdot 7 \cdot 13 \cdot 23 \cdot 29 \cdot 31 \cdot 37 \cdot 41$$
$$538\,257\,874\,440 = 2^3 \cdot 3 \cdot 5 \cdot 11 \cdot 13 \cdot 23 \cdot 29 \cdot 31 \cdot 37 \cdot 41$$

Row 43

$$43 \text{ is Prime}$$
$$903 = 3 \cdot 7 \cdot 43$$
$$12\,341 = 7 \cdot 41 \cdot 43$$
$$123\,410 = 2 \cdot 5 \cdot 7 \cdot 41 \cdot 43$$
$$962\,598 = 2 \cdot 3 \cdot 7 \cdot 13 \cdot 41 \cdot 43$$
$$6\,096\,454 = 2 \cdot 7 \cdot 13 \cdot 19 \cdot 41 \cdot 43$$
$$32\,224\,114 = 2 \cdot 13 \cdot 19 \cdot 37 \cdot 41 \cdot 43$$
$$145\,008\,513 = 3^2 \cdot 13 \cdot 19 \cdot 37 \cdot 41 \cdot 43$$
$$563\,921\,995 = 5 \cdot 7 \cdot 13 \cdot 19 \cdot 37 \cdot 41 \cdot 43$$
$$1\,917\,334\,783 = 7 \cdot 13 \cdot 17 \cdot 19 \cdot 37 \cdot 41 \cdot 43$$
$$5\,752\,004\,349 = 3 \cdot 7 \cdot 13 \cdot 17 \cdot 19 \cdot 37 \cdot 41 \cdot 43$$
$$15\,338\,678\,264 = 2^3 \cdot 7 \cdot 13 \cdot 17 \cdot 19 \cdot 37 \cdot 41 \cdot 43$$
$$36\,576\,848\,168 = 2^3 \cdot 7 \cdot 17 \cdot 19 \cdot 31 \cdot 37 \cdot 41 \cdot 43$$
$$78\,378\,960\,360 = 2^3 \cdot 3 \cdot 5 \cdot 17 \cdot 19 \cdot 31 \cdot 37 \cdot 41 \cdot 43$$
$$151\,532\,656\,696 = 2^3 \cdot 17 \cdot 19 \cdot 29 \cdot 31 \cdot 37 \cdot 41 \cdot 43$$
$$265\,182\,149\,218 = 2 \cdot 7 \cdot 17 \cdot 19 \cdot 29 \cdot 31 \cdot 37 \cdot 41 \cdot 43$$
$$421\,171\,648\,758 = 2 \cdot 3^3 \cdot 7 \cdot 19 \cdot 29 \cdot 31 \cdot 37 \cdot 41 \cdot 43$$
$$608\,359\,048\,206 = 2 \cdot 3 \cdot 7 \cdot 13 \cdot 19 \cdot 29 \cdot 31 \cdot 37 \cdot 41 \cdot 43$$
$$800\,472\,431\,850 = 2 \cdot 3 \cdot 5^2 \cdot 7 \cdot 13 \cdot 29 \cdot 31 \cdot 37 \cdot 41 \cdot 43$$
$$960\,566\,918\,220 = 2^2 \cdot 3^2 \cdot 5 \cdot 7 \cdot 13 \cdot 29 \cdot 31 \cdot 37 \cdot 41 \cdot 43$$
$$1\,052\,049\,481\,860 = 2^2 \cdot 3 \cdot 5 \cdot 13 \cdot 23 \cdot 29 \cdot 31 \cdot 37 \cdot 41 \cdot 43$$

Row 44

$$44 = 2^2 \cdot 11$$
$$946 = 2 \cdot 11 \cdot 43$$
$$13\,244 = 2^2 \cdot 7 \cdot 11 \cdot 43$$
$$135\,751 = 7 \cdot 11 \cdot 41 \cdot 43$$
$$1\,086\,008 = 2^3 \cdot 7 \cdot 11 \cdot 41 \cdot 43$$
$$7\,059\,052 = 2^2 \cdot 7 \cdot 11 \cdot 13 \cdot 41 \cdot 43$$
$$38\,320\,568 = 2^3 \cdot 11 \cdot 13 \cdot 19 \cdot 41 \cdot 43$$
$$177\,232\,627 = 11 \cdot 13 \cdot 19 \cdot 37 \cdot 41 \cdot 43$$
$$708\,930\,508 = 2^2 \cdot 11 \cdot 13 \cdot 19 \cdot 37 \cdot 41 \cdot 43$$
$$2\,481\,256\,778 = 2 \cdot 7 \cdot 11 \cdot 13 \cdot 19 \cdot 37 \cdot 41 \cdot 43$$
$$7\,669\,339\,132 = 2^2 \cdot 7 \cdot 13 \cdot 17 \cdot 19 \cdot 37 \cdot 41 \cdot 43$$
$$21\,090\,682\,613 = 7 \cdot 11 \cdot 13 \cdot 17 \cdot 19 \cdot 37 \cdot 41 \cdot 43$$
$$51\,915\,526\,432 = 2^5 \cdot 7 \cdot 11 \cdot 17 \cdot 19 \cdot 37 \cdot 41 \cdot 43$$
$$114\,955\,808\,528 = 2^4 \cdot 11 \cdot 17 \cdot 19 \cdot 31 \cdot 37 \cdot 41 \cdot 43$$
$$229\,911\,617\,056 = 2^5 \cdot 11 \cdot 17 \cdot 19 \cdot 31 \cdot 37 \cdot 41 \cdot 43$$
$$416\,714\,805\,914 = 2 \cdot 11 \cdot 17 \cdot 19 \cdot 29 \cdot 31 \cdot 37 \cdot 41 \cdot 43$$
$$686\,353\,797\,976 = 2^3 \cdot 7 \cdot 11 \cdot 19 \cdot 29 \cdot 31 \cdot 37 \cdot 41 \cdot 43$$
$$1\,029\,530\,696\,964 = 2^2 \cdot 3 \cdot 7 \cdot 11 \cdot 19 \cdot 29 \cdot 31 \cdot 37 \cdot 41 \cdot 43$$
$$1\,408\,831\,480\,056 = 2^3 \cdot 3 \cdot 7 \cdot 11 \cdot 13 \cdot 29 \cdot 31 \cdot 37 \cdot 41 \cdot 43$$
$$1\,761\,039\,350\,070 = 2 \cdot 3 \cdot 5 \cdot 7 \cdot 11 \cdot 13 \cdot 29 \cdot 31 \cdot 37 \cdot 41 \cdot 43$$
$$2\,012\,616\,400\,080 = 2^4 \cdot 3 \cdot 5 \cdot 11 \cdot 13 \cdot 29 \cdot 31 \cdot 37 \cdot 41 \cdot 43$$
$$2\,104\,098\,963\,720 = 2^3 \cdot 3 \cdot 5 \cdot 13 \cdot 23 \cdot 29 \cdot 31 \cdot 37 \cdot 41 \cdot 43$$

Row 45

$$45 = 3^2 \cdot 5$$
$$990 = 2 \cdot 3^2 \cdot 5 \cdot 11$$
$$14\,190 = 2 \cdot 3 \cdot 5 \cdot 11 \cdot 43$$
$$148\,995 = 3^2 \cdot 5 \cdot 7 \cdot 11 \cdot 43$$
$$1\,221\,759 = 3^2 \cdot 7 \cdot 11 \cdot 41 \cdot 43$$
$$8\,145\,060 = 2^2 \cdot 3 \cdot 5 \cdot 7 \cdot 11 \cdot 41 \cdot 43$$
$$45\,379\,620 = 2^2 \cdot 3^2 \cdot 5 \cdot 11 \cdot 13 \cdot 41 \cdot 43$$
$$215\,553\,195 = 3^2 \cdot 5 \cdot 11 \cdot 13 \cdot 19 \cdot 41 \cdot 43$$
$$886\,163\,135 = 5 \cdot 11 \cdot 13 \cdot 19 \cdot 37 \cdot 41 \cdot 43$$
$$3\,190\,187\,286 = 2 \cdot 3^2 \cdot 11 \cdot 13 \cdot 19 \cdot 37 \cdot 41 \cdot 43$$
$$10\,150\,595\,910 = 2 \cdot 3^2 \cdot 5 \cdot 7 \cdot 13 \cdot 19 \cdot 37 \cdot 41 \cdot 43$$
$$28\,760\,021\,745 = 3 \cdot 5 \cdot 7 \cdot 13 \cdot 17 \cdot 19 \cdot 37 \cdot 41 \cdot 43$$
$$73\,006\,209\,045 = 3^2 \cdot 5 \cdot 7 \cdot 11 \cdot 17 \cdot 19 \cdot 37 \cdot 41 \cdot 43$$
$$166\,871\,334\,960 = 2^4 \cdot 3^2 \cdot 5 \cdot 11 \cdot 17 \cdot 19 \cdot 37 \cdot 41 \cdot 43$$
$$344\,867\,425\,584 = 2^4 \cdot 3 \cdot 11 \cdot 17 \cdot 19 \cdot 31 \cdot 37 \cdot 41 \cdot 43$$
$$646\,626\,422\,970 = 2 \cdot 3^2 \cdot 5 \cdot 11 \cdot 17 \cdot 19 \cdot 31 \cdot 37 \cdot 41 \cdot 43$$
$$1\,103\,068\,603\,890 = 2 \cdot 3^2 \cdot 5 \cdot 11 \cdot 19 \cdot 29 \cdot 31 \cdot 37 \cdot 41 \cdot 43$$
$$1\,715\,884\,494\,940 = 2^2 \cdot 5 \cdot 7 \cdot 11 \cdot 19 \cdot 29 \cdot 31 \cdot 37 \cdot 41 \cdot 43$$
$$2\,438\,362\,177\,020 = 2^2 \cdot 3^3 \cdot 5 \cdot 7 \cdot 11 \cdot 29 \cdot 31 \cdot 37 \cdot 41 \cdot 43$$
$$3\,169\,870\,830\,126 = 2 \cdot 3^3 \cdot 7 \cdot 11 \cdot 13 \cdot 29 \cdot 31 \cdot 37 \cdot 41 \cdot 43$$
$$3\,773\,655\,750\,150 = 2 \cdot 3^2 \cdot 5^2 \cdot 11 \cdot 13 \cdot 29 \cdot 31 \cdot 37 \cdot 41 \cdot 43$$
$$4\,116\,715\,363\,800 = 2^3 \cdot 3^3 \cdot 5^2 \cdot 13 \cdot 29 \cdot 31 \cdot 37 \cdot 41 \cdot 43$$

<u>Row 46</u>

$$46 = 2 \cdot 23$$
$$1\,035 = 3^2 \cdot 5 \cdot 23$$
$$15\,180 = 2^2 \cdot 3 \cdot 5 \cdot 11 \cdot 23$$
$$163\,185 = 3 \cdot 5 \cdot 11 \cdot 23 \cdot 43$$
$$1\,370\,754 = 2 \cdot 3^2 \cdot 7 \cdot 11 \cdot 23 \cdot 43$$
$$9\,366\,819 = 3 \cdot 7 \cdot 11 \cdot 23 \cdot 41 \cdot 43$$
$$53\,524\,680 = 2^3 \cdot 3 \cdot 5 \cdot 11 \cdot 23 \cdot 41 \cdot 43$$
$$260\,932\,815 = 3^2 \cdot 5 \cdot 11 \cdot 13 \cdot 23 \cdot 41 \cdot 43$$
$$1\,101\,716\,330 = 2 \cdot 5 \cdot 11 \cdot 13 \cdot 19 \cdot 23 \cdot 41 \cdot 43$$
$$4\,076\,350\,421 = 11 \cdot 13 \cdot 19 \cdot 23 \cdot 37 \cdot 41 \cdot 43$$
$$13\,340\,783\,196 = 2^2 \cdot 3^2 \cdot 13 \cdot 19 \cdot 23 \cdot 37 \cdot 41 \cdot 43$$
$$38\,910\,617\,655 = 3 \cdot 5 \cdot 7 \cdot 13 \cdot 19 \cdot 23 \cdot 37 \cdot 41 \cdot 43$$
$$101\,766\,230\,790 = 2 \cdot 3 \cdot 5 \cdot 7 \cdot 17 \cdot 19 \cdot 23 \cdot 37 \cdot 41 \cdot 43$$
$$239\,877\,544\,005 = 3^2 \cdot 5 \cdot 11 \cdot 17 \cdot 19 \cdot 23 \cdot 37 \cdot 41 \cdot 43$$
$$511\,738\,760\,544 = 2^5 \cdot 3 \cdot 11 \cdot 17 \cdot 19 \cdot 23 \cdot 37 \cdot 41 \cdot 43$$
$$991\,493\,848\,554 = 2 \cdot 3 \cdot 11 \cdot 17 \cdot 19 \cdot 23 \cdot 31 \cdot 37 \cdot 41 \cdot 43$$
$$1\,749\,695\,026\,860 = 2^2 \cdot 3^2 \cdot 5 \cdot 11 \cdot 19 \cdot 23 \cdot 31 \cdot 37 \cdot 41 \cdot 43$$
$$2\,818\,953\,098\,830 = 2 \cdot 5 \cdot 11 \cdot 19 \cdot 23 \cdot 29 \cdot 31 \cdot 37 \cdot 41 \cdot 43$$
$$4\,154\,246\,671\,960 = 2^3 \cdot 5 \cdot 7 \cdot 11 \cdot 23 \cdot 29 \cdot 31 \cdot 37 \cdot 41 \cdot 43$$
$$5\,608\,233\,007\,146 = 2 \cdot 3^3 \cdot 7 \cdot 11 \cdot 23 \cdot 29 \cdot 31 \cdot 37 \cdot 41 \cdot 43$$
$$6\,943\,526\,580\,276 = 2^2 \cdot 3^2 \cdot 11 \cdot 13 \cdot 23 \cdot 29 \cdot 31 \cdot 37 \cdot 41 \cdot 43$$
$$7\,890\,371\,113\,950 = 2 \cdot 3^2 \cdot 5^2 \cdot 13 \cdot 23 \cdot 29 \cdot 31 \cdot 37 \cdot 41 \cdot 43$$
$$8\,233\,430\,727\,600 = 2^4 \cdot 3^3 \cdot 5^2 \cdot 13 \cdot 29 \cdot 31 \cdot 37 \cdot 41 \cdot 43$$

<u>Row 47</u>

$$47 \text{ is Prime}$$
$$1\,081 = 23 \cdot 47$$
$$16\,215 = 3 \cdot 5 \cdot 23 \cdot 47$$
$$178\,365 = 3 \cdot 5 \cdot 11 \cdot 23 \cdot 47$$
$$1\,533\,939 = 3 \cdot 11 \cdot 23 \cdot 43 \cdot 47$$
$$10\,737\,573 = 3 \cdot 7 \cdot 11 \cdot 23 \cdot 43 \cdot 47$$
$$62\,891\,499 = 3 \cdot 11 \cdot 23 \cdot 41 \cdot 43 \cdot 47$$
$$314\,457\,495 = 3 \cdot 5 \cdot 11 \cdot 23 \cdot 41 \cdot 43 \cdot 47$$
$$1\,362\,649\,145 = 5 \cdot 11 \cdot 13 \cdot 23 \cdot 41 \cdot 43 \cdot 47$$
$$5\,178\,066\,751 = 11 \cdot 13 \cdot 19 \cdot 23 \cdot 41 \cdot 43 \cdot 47$$
$$17\,417\,133\,617 = 13 \cdot 19 \cdot 23 \cdot 37 \cdot 41 \cdot 43 \cdot 47$$
$$52\,251\,400\,851 = 3 \cdot 13 \cdot 19 \cdot 23 \cdot 37 \cdot 41 \cdot 43 \cdot 47$$
$$140\,676\,848\,445 = 3 \cdot 5 \cdot 7 \cdot 19 \cdot 23 \cdot 37 \cdot 41 \cdot 43 \cdot 47$$
$$341\,643\,774\,795 = 3 \cdot 5 \cdot 17 \cdot 19 \cdot 23 \cdot 37 \cdot 41 \cdot 43 \cdot 47$$
$$751\,616\,304\,549 = 3 \cdot 11 \cdot 17 \cdot 19 \cdot 23 \cdot 37 \cdot 41 \cdot 43 \cdot 47$$
$$1\,503\,232\,609\,098 = 2 \cdot 3 \cdot 11 \cdot 17 \cdot 19 \cdot 23 \cdot 37 \cdot 41 \cdot 43 \cdot 47$$
$$2\,741\,188\,875\,414 = 2 \cdot 3 \cdot 11 \cdot 19 \cdot 23 \cdot 31 \cdot 37 \cdot 41 \cdot 43 \cdot 47$$
$$4\,568\,648\,125\,690 = 2 \cdot 5 \cdot 11 \cdot 19 \cdot 23 \cdot 31 \cdot 37 \cdot 41 \cdot 43 \cdot 47$$
$$6\,973\,199\,770\,790 = 2 \cdot 5 \cdot 11 \cdot 23 \cdot 29 \cdot 31 \cdot 37 \cdot 41 \cdot 43 \cdot 47$$
$$9\,762\,479\,679\,106 = 2 \cdot 7 \cdot 11 \cdot 23 \cdot 29 \cdot 31 \cdot 37 \cdot 41 \cdot 43 \cdot 47$$
$$12\,551\,759\,587\,422 = 2 \cdot 3^2 \cdot 11 \cdot 23 \cdot 29 \cdot 31 \cdot 37 \cdot 41 \cdot 43 \cdot 47$$
$$14\,833\,897\,694\,226 = 2 \cdot 3^2 \cdot 13 \cdot 23 \cdot 29 \cdot 31 \cdot 37 \cdot 41 \cdot 43 \cdot 47$$
$$16\,123\,801\,841\,550 = 2 \cdot 3^2 \cdot 5^2 \cdot 13 \cdot 29 \cdot 31 \cdot 37 \cdot 41 \cdot 43 \cdot 47$$

Row 48

48	=	$2^4 \cdot 3$
1 128	=	$2^3 \cdot 3 \cdot 47$
17 296	=	$2^4 \cdot 23 \cdot 47$
194 580	=	$2^2 \cdot 3^2 \cdot 5 \cdot 23 \cdot 47$
1 712 304	=	$2^4 \cdot 3^2 \cdot 11 \cdot 23 \cdot 47$
12 271 512	=	$2^3 \cdot 3 \cdot 11 \cdot 23 \cdot 43 \cdot 47$
73 629 072	=	$2^4 \cdot 3^2 \cdot 11 \cdot 23 \cdot 43 \cdot 47$
377 348 994	=	$2 \cdot 3^2 \cdot 11 \cdot 23 \cdot 41 \cdot 43 \cdot 47$
1 677 106 640	=	$2^4 \cdot 5 \cdot 11 \cdot 23 \cdot 41 \cdot 43 \cdot 47$
6 540 715 896	=	$2^3 \cdot 3 \cdot 11 \cdot 13 \cdot 23 \cdot 41 \cdot 43 \cdot 47$
22 595 200 368	=	$2^4 \cdot 3 \cdot 13 \cdot 19 \cdot 23 \cdot 41 \cdot 43 \cdot 47$
69 668 534 468	=	$2^2 \cdot 13 \cdot 19 \cdot 23 \cdot 37 \cdot 41 \cdot 43 \cdot 47$
192 928 249 296	=	$2^4 \cdot 3^2 \cdot 19 \cdot 23 \cdot 37 \cdot 41 \cdot 43 \cdot 47$
482 320 623 240	=	$2^3 \cdot 3^2 \cdot 5 \cdot 19 \cdot 23 \cdot 37 \cdot 41 \cdot 43 \cdot 47$
1 093 260 079 344	=	$2^4 \cdot 3 \cdot 17 \cdot 19 \cdot 23 \cdot 37 \cdot 41 \cdot 43 \cdot 47$
2 254 848 913 647	=	$3^2 \cdot 11 \cdot 17 \cdot 19 \cdot 23 \cdot 37 \cdot 41 \cdot 43 \cdot 47$
4 244 421 484 512	=	$2^5 \cdot 3^2 \cdot 11 \cdot 19 \cdot 23 \cdot 37 \cdot 41 \cdot 43 \cdot 47$
7 309 837 001 104	=	$2^4 \cdot 11 \cdot 19 \cdot 23 \cdot 31 \cdot 37 \cdot 41 \cdot 43 \cdot 47$
11 541 847 896 480	=	$2^5 \cdot 3 \cdot 5 \cdot 11 \cdot 23 \cdot 31 \cdot 37 \cdot 41 \cdot 43 \cdot 47$
16 735 679 449 896	=	$2^3 \cdot 3 \cdot 11 \cdot 23 \cdot 29 \cdot 31 \cdot 37 \cdot 41 \cdot 43 \cdot 47$
22 314 239 266 528	=	$2^5 \cdot 11 \cdot 23 \cdot 29 \cdot 31 \cdot 37 \cdot 41 \cdot 43 \cdot 47$
27 385 657 281 648	=	$2^4 \cdot 3^3 \cdot 23 \cdot 29 \cdot 31 \cdot 37 \cdot 41 \cdot 43 \cdot 47$
30 957 699 535 776	=	$2^5 \cdot 3^3 \cdot 13 \cdot 29 \cdot 31 \cdot 37 \cdot 41 \cdot 43 \cdot 47$
32 247 603 683 100	=	$2^2 \cdot 3^2 \cdot 5^2 \cdot 13 \cdot 29 \cdot 31 \cdot 37 \cdot 41 \cdot 43 \cdot 47$

Row 49

49	=	7^2
1 176	=	$2^3 \cdot 3 \cdot 7^2$
18 424	=	$2^3 \cdot 7^2 \cdot 47$
211 876	=	$2^2 \cdot 7^2 \cdot 23 \cdot 47$
1 906 884	=	$2^2 \cdot 3^2 \cdot 7^2 \cdot 23 \cdot 47$
13 983 816	=	$2^3 \cdot 3 \cdot 7^2 \cdot 11 \cdot 23 \cdot 47$
85 900 584	=	$2^3 \cdot 3 \cdot 7 \cdot 11 \cdot 23 \cdot 43 \cdot 47$
450 978 066	=	$2 \cdot 3^2 \cdot 7^2 \cdot 11 \cdot 23 \cdot 43 \cdot 47$
2 054 455 634	=	$2 \cdot 7^2 \cdot 11 \cdot 23 \cdot 41 \cdot 43 \cdot 47$
8 217 822 536	=	$2^3 \cdot 7^2 \cdot 11 \cdot 23 \cdot 41 \cdot 43 \cdot 47$
29 135 916 264	=	$2^3 \cdot 3 \cdot 7^2 \cdot 13 \cdot 23 \cdot 41 \cdot 43 \cdot 47$
92 263 734 836	=	$2^2 \cdot 7^2 \cdot 13 \cdot 19 \cdot 23 \cdot 41 \cdot 43 \cdot 47$
262 596 783 764	=	$2^2 \cdot 7^2 \cdot 19 \cdot 23 \cdot 37 \cdot 41 \cdot 43 \cdot 47$
675 248 872 536	=	$2^3 \cdot 3^2 \cdot 7 \cdot 19 \cdot 23 \cdot 37 \cdot 41 \cdot 43 \cdot 47$
1 575 580 702 584	=	$2^3 \cdot 3 \cdot 7^2 \cdot 19 \cdot 23 \cdot 37 \cdot 41 \cdot 43 \cdot 47$
3 348 108 992 991	=	$3 \cdot 7^2 \cdot 17 \cdot 19 \cdot 23 \cdot 37 \cdot 41 \cdot 43 \cdot 47$
6 499 270 398 159	=	$3^2 \cdot 7^2 \cdot 11 \cdot 19 \cdot 23 \cdot 37 \cdot 41 \cdot 43 \cdot 47$
11 554 258 485 616	=	$2^4 \cdot 7^2 \cdot 11 \cdot 19 \cdot 23 \cdot 37 \cdot 41 \cdot 43 \cdot 47$
18 851 684 897 584	=	$2^4 \cdot 7^2 \cdot 11 \cdot 23 \cdot 31 \cdot 37 \cdot 41 \cdot 43 \cdot 47$
28 277 527 346 376	=	$2^3 \cdot 3 \cdot 7^2 \cdot 11 \cdot 23 \cdot 31 \cdot 37 \cdot 41 \cdot 43 \cdot 47$
39 049 918 716 424	=	$2^3 \cdot 7 \cdot 11 \cdot 23 \cdot 29 \cdot 31 \cdot 37 \cdot 41 \cdot 43 \cdot 47$
40 600 806 548 176	=	$2^4 \cdot 7^2 \cdot 23 \cdot 29 \cdot 31 \cdot 37 \cdot 41 \cdot 43 \cdot 47$
58 343 356 817 424	=	$2^4 \cdot 3^3 \cdot 7^2 \cdot 29 \cdot 31 \cdot 37 \cdot 41 \cdot 43 \cdot 47$
63 205 303 218 876	=	$2^2 \cdot 3^2 \cdot 7^2 \cdot 13 \cdot 29 \cdot 31 \cdot 37 \cdot 41 \cdot 43 \cdot 47$

Row 50

$$50 = 2 \cdot 5^2$$
$$1\,225 = 5^2 \cdot 7^2$$
$$19\,600 = 2^4 \cdot 5^2 \cdot 7^2$$
$$230\,300 = 2^2 \cdot 5^2 \cdot 7^2 \cdot 47$$
$$2\,118\,760 = 2^3 \cdot 5 \cdot 7^2 \cdot 23 \cdot 47$$
$$15\,890\,700 = 2^2 \cdot 3 \cdot 5^2 \cdot 7^2 \cdot 23 \cdot 47$$
$$99\,884\,400 = 2^4 \cdot 3 \cdot 5^2 \cdot 7 \cdot 11 \cdot 23 \cdot 47$$
$$536\,878\,650 = 2 \cdot 3 \cdot 5^2 \cdot 7 \cdot 11 \cdot 23 \cdot 43 \cdot 47$$
$$2\,505\,433\,700 = 2^2 \cdot 5^2 \cdot 7^2 \cdot 11 \cdot 23 \cdot 43 \cdot 47$$
$$10\,272\,278\,170 = 2 \cdot 5 \cdot 7^2 \cdot 11 \cdot 23 \cdot 41 \cdot 43 \cdot 47$$
$$37\,353\,738\,800 = 2^4 \cdot 5^2 \cdot 7^2 \cdot 23 \cdot 41 \cdot 43 \cdot 47$$
$$121\,399\,651\,100 = 2^2 \cdot 5^2 \cdot 7^2 \cdot 13 \cdot 23 \cdot 41 \cdot 43 \cdot 47$$
$$354\,860\,518\,600 = 2^3 \cdot 5^2 \cdot 7^2 \cdot 19 \cdot 23 \cdot 41 \cdot 43 \cdot 47$$
$$937\,845\,656\,300 = 2^2 \cdot 5^2 \cdot 7 \cdot 19 \cdot 23 \cdot 37 \cdot 41 \cdot 43 \cdot 47$$
$$2\,250\,829\,575\,120 = 2^4 \cdot 3 \cdot 5 \cdot 7 \cdot 19 \cdot 23 \cdot 37 \cdot 41 \cdot 43 \cdot 47$$
$$4\,923\,689\,695\,575 = 3 \cdot 5^2 \cdot 7^2 \cdot 19 \cdot 23 \cdot 37 \cdot 41 \cdot 43 \cdot 47$$
$$9\,847\,379\,391\,150 = 2 \cdot 3 \cdot 5^2 \cdot 7^2 \cdot 19 \cdot 23 \cdot 37 \cdot 41 \cdot 43 \cdot 47$$
$$18\,053\,528\,883\,775 = 5^2 \cdot 7^2 \cdot 11 \cdot 19 \cdot 23 \cdot 37 \cdot 41 \cdot 43 \cdot 47$$
$$30\,405\,943\,383\,200 = 2^5 \cdot 5^2 \cdot 7^2 \cdot 11 \cdot 23 \cdot 37 \cdot 41 \cdot 43 \cdot 47$$
$$47\,129\,212\,243\,960 = 2^3 \cdot 5 \cdot 7^2 \cdot 11 \cdot 23 \cdot 31 \cdot 37 \cdot 41 \cdot 43 \cdot 47$$
$$67\,327\,446\,062\,800 = 2^4 \cdot 5^2 \cdot 7 \cdot 11 \cdot 23 \cdot 31 \cdot 37 \cdot 41 \cdot 43 \cdot 47$$
$$88\,749\,815\,264\,600 = 2^3 \cdot 5^2 \cdot 7 \cdot 23 \cdot 29 \cdot 31 \cdot 37 \cdot 41 \cdot 43 \cdot 47$$
$$108\,043\,253\,365\,600 = 2^5 \cdot 5^2 \cdot 7^2 \cdot 29 \cdot 31 \cdot 37 \cdot 41 \cdot 43 \cdot 47$$
$$121\,548\,660\,036\,300 = 2^2 \cdot 3^2 \cdot 5^2 \cdot 7^2 \cdot 29 \cdot 31 \cdot 37 \cdot 41 \cdot 43 \cdot 47$$
$$126\,410\,606\,437\,752 = 2^3 \cdot 3^2 \cdot 7^2 \cdot 13 \cdot 29 \cdot 31 \cdot 37 \cdot 41 \cdot 43 \cdot 47$$

Row 51

$$51 = 3 \cdot 17$$
$$1\,275 = 3 \cdot 5^2 \cdot 17$$
$$20\,825 = 5^2 \cdot 7^2 \cdot 17$$
$$249\,900 = 2^2 \cdot 3 \cdot 5^2 \cdot 7^2 \cdot 17$$
$$2\,349\,060 = 2^2 \cdot 3 \cdot 5 \cdot 7^2 \cdot 17 \cdot 47$$
$$18\,009\,460 = 2^2 \cdot 5 \cdot 7^2 \cdot 17 \cdot 23 \cdot 47$$
$$115\,775\,100 = 2^2 \cdot 3^2 \cdot 5^2 \cdot 7 \cdot 17 \cdot 23 \cdot 47$$
$$636\,763\,050 = 2 \cdot 3^2 \cdot 5^2 \cdot 7 \cdot 11 \cdot 17 \cdot 23 \cdot 47$$
$$3\,042\,312\,350 = 2 \cdot 5^2 \cdot 7 \cdot 11 \cdot 17 \cdot 23 \cdot 43 \cdot 47$$
$$12\,777\,711\,870 = 2 \cdot 3 \cdot 5 \cdot 7^2 \cdot 11 \cdot 17 \cdot 23 \cdot 43 \cdot 47$$
$$47\,626\,016\,970 = 2 \cdot 3 \cdot 5 \cdot 7^2 \cdot 17 \cdot 23 \cdot 41 \cdot 43 \cdot 47$$
$$158\,753\,389\,900 = 2^2 \cdot 5^2 \cdot 7^2 \cdot 17 \cdot 23 \cdot 41 \cdot 43 \cdot 47$$
$$476\,260\,169\,700 = 2^2 \cdot 3 \cdot 5^2 \cdot 7^2 \cdot 17 \cdot 23 \cdot 41 \cdot 43 \cdot 47$$
$$1\,292\,706\,174\,900 = 2^2 \cdot 3 \cdot 5^2 \cdot 7 \cdot 17 \cdot 19 \cdot 23 \cdot 41 \cdot 43 \cdot 47$$
$$3\,188\,675\,231\,420 = 2^2 \cdot 5 \cdot 7 \cdot 17 \cdot 19 \cdot 23 \cdot 37 \cdot 41 \cdot 43 \cdot 47$$
$$7\,174\,519\,270\,695 = 3^2 \cdot 5 \cdot 7 \cdot 17 \cdot 19 \cdot 23 \cdot 37 \cdot 41 \cdot 43 \cdot 47$$
$$14\,771\,069\,086\,725 = 3^2 \cdot 5^2 \cdot 7^2 \cdot 19 \cdot 23 \cdot 37 \cdot 41 \cdot 43 \cdot 47$$
$$27\,900\,908\,274\,925 = 5^2 \cdot 7^2 \cdot 17 \cdot 19 \cdot 23 \cdot 37 \cdot 41 \cdot 43 \cdot 47$$
$$48\,459\,472\,266\,975 = 3 \cdot 5^2 \cdot 7^2 \cdot 11 \cdot 17 \cdot 23 \cdot 37 \cdot 41 \cdot 43 \cdot 47$$
$$77\,535\,155\,627\,160 = 2^3 \cdot 3 \cdot 5 \cdot 7^2 \cdot 11 \cdot 17 \cdot 23 \cdot 37 \cdot 41 \cdot 43 \cdot 47$$
$$114\,456\,658\,306\,760 = 2^3 \cdot 5 \cdot 7 \cdot 11 \cdot 17 \cdot 23 \cdot 31 \cdot 37 \cdot 41 \cdot 43 \cdot 47$$
$$156\,077\,261\,327\,400 = 2^3 \cdot 3 \cdot 5^2 \cdot 7 \cdot 17 \cdot 23 \cdot 31 \cdot 37 \cdot 41 \cdot 43 \cdot 47$$
$$196\,793\,068\,630\,200 = 2^3 \cdot 3 \cdot 5^2 \cdot 7 \cdot 17 \cdot 29 \cdot 31 \cdot 37 \cdot 41 \cdot 43 \cdot 47$$
$$229\,591\,913\,401\,900 = 2^2 \cdot 5^2 \cdot 7^2 \cdot 17 \cdot 29 \cdot 31 \cdot 37 \cdot 41 \cdot 43 \cdot 47$$
$$247\,959\,266\,474\,052 = 2^2 \cdot 3^3 \cdot 7^2 \cdot 17 \cdot 29 \cdot 31 \cdot 37 \cdot 41 \cdot 43 \cdot 47$$

<u>Row 52</u>

$52 = 2^2 \cdot 13$

$1\,326 = 2 \cdot 3 \cdot 13 \cdot 17$

$22\,100 = 2^2 \cdot 5^2 \cdot 13 \cdot 17$

$270\,725 = 5^2 \cdot 7^2 \cdot 13 \cdot 17$

$2\,598\,960 = 2^4 \cdot 3 \cdot 5 \cdot 7^2 \cdot 13 \cdot 17$

$20\,358\,520 = 2^3 \cdot 5 \cdot 7^2 \cdot 13 \cdot 17 \cdot 47$

$133\,784\,560 = 2^4 \cdot 5 \cdot 7 \cdot 13 \cdot 17 \cdot 23 \cdot 47$

$752\,538\,150 = 2 \cdot 3^2 \cdot 5^2 \cdot 7 \cdot 13 \cdot 17 \cdot 23 \cdot 47$

$3\,679\,075\,400 = 2^3 \cdot 5^2 \cdot 7 \cdot 11 \cdot 13 \cdot 17 \cdot 23 \cdot 47$

$15\,820\,024\,220 = 2^2 \cdot 5 \cdot 7 \cdot 11 \cdot 13 \cdot 17 \cdot 23 \cdot 43 \cdot 47$

$60\,403\,728\,840 = 2^3 \cdot 3 \cdot 5 \cdot 7^2 \cdot 13 \cdot 17 \cdot 23 \cdot 43 \cdot 47$

$206\,379\,406\,870 = 2 \cdot 5 \cdot 7^2 \cdot 13 \cdot 17 \cdot 23 \cdot 41 \cdot 43 \cdot 47$

$635\,013\,559\,600 = 2^4 \cdot 5^2 \cdot 7^2 \cdot 17 \cdot 23 \cdot 41 \cdot 43 \cdot 47$

$1\,768\,966\,344\,600 = 2^3 \cdot 3 \cdot 5^2 \cdot 7 \cdot 13 \cdot 17 \cdot 23 \cdot 41 \cdot 43 \cdot 47$

$4\,481\,381\,406\,320 = 2^4 \cdot 5 \cdot 7 \cdot 13 \cdot 17 \cdot 19 \cdot 23 \cdot 41 \cdot 43 \cdot 47$

$10\,363\,194\,502\,115 = 5 \cdot 7 \cdot 13 \cdot 17 \cdot 19 \cdot 23 \cdot 37 \cdot 41 \cdot 43 \cdot 47$

$21\,945\,588\,357\,420 = 2^2 \cdot 3^2 \cdot 5 \cdot 7 \cdot 13 \cdot 19 \cdot 23 \cdot 37 \cdot 41 \cdot 43 \cdot 47$

$42\,671\,977\,361\,650 = 2 \cdot 5^2 \cdot 7^2 \cdot 13 \cdot 19 \cdot 23 \cdot 37 \cdot 41 \cdot 43 \cdot 47$

$76\,360\,380\,541\,900 = 2^2 \cdot 5^2 \cdot 7^2 \cdot 13 \cdot 17 \cdot 23 \cdot 37 \cdot 41 \cdot 43 \cdot 47$

$125\,994\,627\,894\,135 = 3 \cdot 5 \cdot 7^2 \cdot 11 \cdot 13 \cdot 17 \cdot 23 \cdot 37 \cdot 41 \cdot 43 \cdot 47$

$191\,991\,813\,933\,920 = 2^5 \cdot 5 \cdot 7 \cdot 11 \cdot 13 \cdot 17 \cdot 23 \cdot 37 \cdot 41 \cdot 43 \cdot 47$

$270\,533\,919\,634\,160 = 2^4 \cdot 5 \cdot 7 \cdot 13 \cdot 17 \cdot 23 \cdot 31 \cdot 37 \cdot 41 \cdot 43 \cdot 47$

$352\,870\,329\,957\,600 = 2^5 \cdot 3 \cdot 5^2 \cdot 7 \cdot 13 \cdot 17 \cdot 31 \cdot 37 \cdot 41 \cdot 43 \cdot 47$

$426\,384\,982\,032\,100 = 2^2 \cdot 5^2 \cdot 7 \cdot 13 \cdot 17 \cdot 29 \cdot 31 \cdot 37 \cdot 41 \cdot 43 \cdot 47$

$477\,551\,179\,875\,952 = 2^4 \cdot 7^2 \cdot 13 \cdot 17 \cdot 29 \cdot 31 \cdot 37 \cdot 41 \cdot 43 \cdot 47$

$495\,918\,532\,948\,104 = 2^3 \cdot 3^3 \cdot 7^2 \cdot 17 \cdot 29 \cdot 31 \cdot 37 \cdot 41 \cdot 43 \cdot 47$

<u>Row 53</u>

$$53 \text{ is Prime}$$

$$1\,378 = 2 \cdot 13 \cdot 53$$

$$23\,426 = 2 \cdot 13 \cdot 17 \cdot 53$$

$$292\,825 = 5^2 \cdot 13 \cdot 17 \cdot 53$$

$$2\,869\,685 = 5 \cdot 7^2 \cdot 13 \cdot 17 \cdot 53$$

$$22\,957\,480 = 2^3 \cdot 5 \cdot 7^2 \cdot 13 \cdot 17 \cdot 53$$

$$154\,143\,080 = 2^3 \cdot 5 \cdot 7 \cdot 13 \cdot 17 \cdot 47 \cdot 53$$

$$886\,322\,710 = 2 \cdot 5 \cdot 7 \cdot 13 \cdot 17 \cdot 23 \cdot 47 \cdot 53$$

$$4\,431\,613\,550 = 2 \cdot 5^2 \cdot 7 \cdot 13 \cdot 17 \cdot 23 \cdot 47 \cdot 53$$

$$19\,499\,099\,620 = 2^2 \cdot 5 \cdot 7 \cdot 11 \cdot 13 \cdot 17 \cdot 23 \cdot 47 \cdot 53$$

$$76\,223\,753\,060 = 2^2 \cdot 5 \cdot 7 \cdot 13 \cdot 17 \cdot 23 \cdot 43 \cdot 47 \cdot 53$$

$$266\,783\,135\,710 = 2 \cdot 5 \cdot 7^2 \cdot 13 \cdot 17 \cdot 23 \cdot 43 \cdot 47 \cdot 53$$

$$841\,392\,966\,470 = 2 \cdot 5 \cdot 7^2 \cdot 17 \cdot 23 \cdot 41 \cdot 43 \cdot 47 \cdot 53$$

$$2\,403\,979\,904\,200 = 2^3 \cdot 5^2 \cdot 7 \cdot 17 \cdot 23 \cdot 41 \cdot 43 \cdot 47 \cdot 53$$

$$6\,250\,347\,750\,920 = 2^3 \cdot 5 \cdot 7 \cdot 13 \cdot 17 \cdot 23 \cdot 41 \cdot 43 \cdot 47 \cdot 53$$

$$14\,844\,575\,908\,435 = 5 \cdot 7 \cdot 13 \cdot 17 \cdot 19 \cdot 23 \cdot 41 \cdot 43 \cdot 47 \cdot 53$$

$$32\,308\,782\,859\,535 = 5 \cdot 7 \cdot 13 \cdot 19 \cdot 23 \cdot 37 \cdot 41 \cdot 43 \cdot 47 \cdot 53$$

$$64\,617\,565\,719\,070 = 2 \cdot 5 \cdot 7 \cdot 13 \cdot 19 \cdot 23 \cdot 37 \cdot 41 \cdot 43 \cdot 47 \cdot 53$$

$$119\,032\,357\,903\,550 = 2 \cdot 5^2 \cdot 7^2 \cdot 13 \cdot 23 \cdot 37 \cdot 41 \cdot 43 \cdot 47 \cdot 53$$

$$202\,355\,008\,436\,035 = 5 \cdot 7^2 \cdot 13 \cdot 17 \cdot 23 \cdot 37 \cdot 41 \cdot 43 \cdot 47 \cdot 53$$

$$317\,986\,441\,828\,055 = 5 \cdot 7 \cdot 11 \cdot 13 \cdot 17 \cdot 23 \cdot 37 \cdot 41 \cdot 43 \cdot 47 \cdot 53$$

$$462\,525\,733\,568\,080 = 2^4 \cdot 5 \cdot 7 \cdot 13 \cdot 17 \cdot 23 \cdot 37 \cdot 41 \cdot 43 \cdot 47 \cdot 53$$

$$623\,404\,249\,591\,760 = 2^4 \cdot 5 \cdot 7 \cdot 13 \cdot 17 \cdot 31 \cdot 37 \cdot 41 \cdot 43 \cdot 47 \cdot 53$$

$$779\,255\,311\,989\,700 = 2^2 \cdot 5^2 \cdot 7 \cdot 13 \cdot 17 \cdot 31 \cdot 37 \cdot 41 \cdot 43 \cdot 47 \cdot 53$$

$$903\,936\,161\,908\,052 = 2^2 \cdot 7 \cdot 13 \cdot 17 \cdot 29 \cdot 31 \cdot 37 \cdot 41 \cdot 43 \cdot 47 \cdot 53$$

$$973\,469\,712\,824\,056 = 2^3 \cdot 7^2 \cdot 17 \cdot 29 \cdot 31 \cdot 37 \cdot 41 \cdot 43 \cdot 47 \cdot 53$$

Row 54

$$54 = 2 \cdot 3^3$$
$$1\,431 = 3^3 \cdot 53$$
$$24\,804 = 2^2 \cdot 3^2 \cdot 13 \cdot 53$$
$$316\,251 = 3^3 \cdot 13 \cdot 17 \cdot 53$$
$$3\,162\,510 = 2 \cdot 3^3 \cdot 5 \cdot 13 \cdot 17 \cdot 53$$
$$25\,827\,165 = 3^2 \cdot 5 \cdot 7^2 \cdot 13 \cdot 17 \cdot 53$$
$$177\,100\,560 = 2^4 \cdot 3^3 \cdot 5 \cdot 7 \cdot 13 \cdot 17 \cdot 53$$
$$1\,040\,465\,790 = 2 \cdot 3^3 \cdot 5 \cdot 7 \cdot 13 \cdot 17 \cdot 47 \cdot 53$$
$$5\,317\,936\,260 = 2^2 \cdot 3 \cdot 5 \cdot 7 \cdot 13 \cdot 17 \cdot 23 \cdot 47 \cdot 53$$
$$23\,930\,713\,170 = 2 \cdot 3^3 \cdot 5 \cdot 7 \cdot 13 \cdot 17 \cdot 23 \cdot 47 \cdot 53$$
$$95\,722\,852\,680 = 2^3 \cdot 3^3 \cdot 5 \cdot 7 \cdot 13 \cdot 17 \cdot 23 \cdot 47 \cdot 53$$
$$343\,006\,888\,770 = 2 \cdot 3^2 \cdot 5 \cdot 7 \cdot 13 \cdot 17 \cdot 23 \cdot 43 \cdot 47 \cdot 53$$
$$1\,108\,176\,102\,180 = 2^2 \cdot 3^3 \cdot 5 \cdot 7^2 \cdot 17 \cdot 23 \cdot 43 \cdot 47 \cdot 53$$
$$3\,245\,372\,870\,670 = 2 \cdot 3^3 \cdot 5 \cdot 7 \cdot 17 \cdot 23 \cdot 41 \cdot 43 \cdot 47 \cdot 53$$
$$8\,654\,327\,655\,120 = 2^4 \cdot 3^2 \cdot 5 \cdot 7 \cdot 17 \cdot 23 \cdot 41 \cdot 43 \cdot 47 \cdot 53$$
$$21\,094\,923\,659\,355 = 3^3 \cdot 5 \cdot 7 \cdot 13 \cdot 17 \cdot 23 \cdot 41 \cdot 43 \cdot 47 \cdot 53$$
$$47\,153\,358\,767\,970 = 2 \cdot 3^3 \cdot 5 \cdot 7 \cdot 13 \cdot 19 \cdot 23 \cdot 41 \cdot 43 \cdot 47 \cdot 53$$
$$96\,926\,348\,578\,605 = 3 \cdot 5 \cdot 7 \cdot 13 \cdot 19 \cdot 23 \cdot 37 \cdot 41 \cdot 43 \cdot 47 \cdot 53$$
$$183\,649\,923\,622\,620 = 2^2 \cdot 3^3 \cdot 5 \cdot 7 \cdot 13 \cdot 23 \cdot 37 \cdot 41 \cdot 43 \cdot 47 \cdot 53$$
$$321\,387\,366\,339\,585 = 3^3 \cdot 5 \cdot 7^2 \cdot 13 \cdot 23 \cdot 37 \cdot 41 \cdot 43 \cdot 47 \cdot 53$$
$$520\,341\,450\,264\,090 = 2 \cdot 3^2 \cdot 5 \cdot 7 \cdot 13 \cdot 17 \cdot 23 \cdot 37 \cdot 41 \cdot 43 \cdot 47 \cdot 53$$
$$780\,512\,175\,396\,135 = 3^3 \cdot 5 \cdot 7 \cdot 13 \cdot 17 \cdot 23 \cdot 37 \cdot 41 \cdot 43 \cdot 47 \cdot 53$$
$$1\,085\,929\,983\,159\,840 = 2^5 \cdot 3^3 \cdot 5 \cdot 7 \cdot 13 \cdot 17 \cdot 37 \cdot 41 \cdot 43 \cdot 47 \cdot 53$$
$$1\,402\,659\,561\,581\,460 = 2^2 \cdot 3^2 \cdot 5 \cdot 7 \cdot 13 \cdot 17 \cdot 31 \cdot 37 \cdot 41 \cdot 43 \cdot 47 \cdot 53$$
$$1\,683\,191\,473\,897\,752 = 2^3 \cdot 3^3 \cdot 7 \cdot 13 \cdot 17 \cdot 31 \cdot 37 \cdot 41 \cdot 43 \cdot 47 \cdot 53$$
$$1\,877\,405\,874\,732\,108 = 2^2 \cdot 3^3 \cdot 7 \cdot 17 \cdot 29 \cdot 31 \cdot 37 \cdot 41 \cdot 43 \cdot 47 \cdot 53$$
$$1\,946\,939\,425\,648\,112 = 2^4 \cdot 7^2 \cdot 17 \cdot 29 \cdot 31 \cdot 37 \cdot 41 \cdot 43 \cdot 47 \cdot 53$$

Row 55

$$55 = 5 \cdot 11$$
$$1\,485 = 3^3 \cdot 5 \cdot 11$$
$$26\,235 = 3^2 \cdot 5 \cdot 11 \cdot 53$$
$$341\,055 = 3^2 \cdot 5 \cdot 11 \cdot 13 \cdot 53$$
$$3\,478\,761 = 3^3 \cdot 11 \cdot 13 \cdot 17 \cdot 53$$
$$28\,989\,675 = 3^2 \cdot 5^2 \cdot 11 \cdot 13 \cdot 17 \cdot 53$$
$$202\,927\,725 = 3^2 \cdot 5^2 \cdot 7 \cdot 11 \cdot 13 \cdot 17 \cdot 53$$
$$1\,217\,566\,350 = 2 \cdot 3^3 \cdot 5^2 \cdot 7 \cdot 11 \cdot 13 \cdot 17 \cdot 53$$
$$6\,358\,402\,050 = 2 \cdot 3 \cdot 5^2 \cdot 7 \cdot 11 \cdot 13 \cdot 17 \cdot 47 \cdot 53$$
$$29\,248\,649\,430 = 2 \cdot 3 \cdot 5 \cdot 7 \cdot 11 \cdot 13 \cdot 17 \cdot 23 \cdot 47 \cdot 53$$
$$119\,653\,565\,850 = 2 \cdot 3^3 \cdot 5^2 \cdot 7 \cdot 13 \cdot 17 \cdot 23 \cdot 47 \cdot 53$$
$$438\,729\,741\,450 = 2 \cdot 3^2 \cdot 5^2 \cdot 7 \cdot 11 \cdot 13 \cdot 17 \cdot 23 \cdot 47 \cdot 53$$
$$1\,451\,182\,990\,950 = 2 \cdot 3^2 \cdot 5^2 \cdot 7 \cdot 11 \cdot 17 \cdot 23 \cdot 43 \cdot 47 \cdot 53$$
$$4\,353\,548\,972\,850 = 2 \cdot 3^3 \cdot 5^2 \cdot 7 \cdot 11 \cdot 17 \cdot 23 \cdot 43 \cdot 47 \cdot 53$$
$$11\,899\,700\,525\,790 = 2 \cdot 3^2 \cdot 5 \cdot 7 \cdot 11 \cdot 17 \cdot 23 \cdot 41 \cdot 43 \cdot 47 \cdot 53$$
$$29\,749\,251\,314\,475 = 3^2 \cdot 5^2 \cdot 7 \cdot 11 \cdot 17 \cdot 23 \cdot 41 \cdot 43 \cdot 47 \cdot 53$$
$$68\,248\,282\,427\,325 = 3^3 \cdot 5^2 \cdot 7 \cdot 11 \cdot 13 \cdot 23 \cdot 41 \cdot 43 \cdot 47 \cdot 53$$
$$144\,079\,707\,346\,575 = 3 \cdot 5^2 \cdot 7 \cdot 11 \cdot 13 \cdot 19 \cdot 23 \cdot 41 \cdot 43 \cdot 47 \cdot 53$$
$$280\,576\,272\,201\,225 = 3 \cdot 5^2 \cdot 7 \cdot 11 \cdot 13 \cdot 23 \cdot 37 \cdot 41 \cdot 43 \cdot 47 \cdot 53$$
$$505\,037\,289\,962\,205 = 3^3 \cdot 5 \cdot 7 \cdot 11 \cdot 13 \cdot 23 \cdot 37 \cdot 41 \cdot 43 \cdot 47 \cdot 53$$
$$841\,728\,816\,603\,675 = 3^2 \cdot 5^2 \cdot 7 \cdot 11 \cdot 13 \cdot 23 \cdot 37 \cdot 41 \cdot 43 \cdot 47 \cdot 53$$
$$1\,300\,853\,625\,660\,225 = 3^2 \cdot 5^2 \cdot 7 \cdot 13 \cdot 17 \cdot 23 \cdot 37 \cdot 41 \cdot 43 \cdot 47 \cdot 53$$
$$1\,866\,442\,158\,555\,975 = 3^3 \cdot 5^2 \cdot 7 \cdot 11 \cdot 13 \cdot 17 \cdot 37 \cdot 41 \cdot 43 \cdot 47 \cdot 53$$
$$2\,488\,589\,544\,741\,300 = 2^2 \cdot 3^2 \cdot 5^2 \cdot 7 \cdot 11 \cdot 13 \cdot 17 \cdot 37 \cdot 41 \cdot 43 \cdot 47 \cdot 53$$
$$3\,085\,851\,035\,479\,212 = 2^2 \cdot 3^2 \cdot 7 \cdot 11 \cdot 13 \cdot 17 \cdot 31 \cdot 37 \cdot 41 \cdot 43 \cdot 47 \cdot 53$$
$$3\,560\,597\,348\,629\,860 = 2^2 \cdot 3^3 \cdot 5 \cdot 7 \cdot 11 \cdot 17 \cdot 31 \cdot 37 \cdot 41 \cdot 43 \cdot 47 \cdot 53$$
$$3\,824\,345\,300\,380\,220 = 2^2 \cdot 5 \cdot 7 \cdot 11 \cdot 17 \cdot 29 \cdot 31 \cdot 37 \cdot 41 \cdot 43 \cdot 47 \cdot 53$$

<u>Row 56</u>

$$56 = 2^3 \cdot 7$$
$$1\,540 = 2^2 \cdot 5 \cdot 7 \cdot 11$$
$$27\,720 = 2^3 \cdot 3^2 \cdot 5 \cdot 7 \cdot 11$$
$$367\,290 = 2 \cdot 3^2 \cdot 5 \cdot 7 \cdot 11 \cdot 53$$
$$3\,819\,816 = 2^3 \cdot 3^2 \cdot 7 \cdot 11 \cdot 13 \cdot 53$$
$$32\,468\,436 = 2^2 \cdot 3^2 \cdot 7 \cdot 11 \cdot 13 \cdot 17 \cdot 53$$
$$231\,917\,400 = 2^3 \cdot 3^2 \cdot 5^2 \cdot 11 \cdot 13 \cdot 17 \cdot 53$$
$$1\,420\,494\,075 = 3^2 \cdot 5^2 \cdot 7^2 \cdot 11 \cdot 13 \cdot 17 \cdot 53$$
$$7\,575\,968\,400 = 2^4 \cdot 3 \cdot 5^2 \cdot 7^2 \cdot 11 \cdot 13 \cdot 17 \cdot 53$$
$$35\,607\,051\,480 = 2^3 \cdot 3 \cdot 5 \cdot 7^2 \cdot 11 \cdot 13 \cdot 17 \cdot 47 \cdot 53$$
$$148\,902\,215\,280 = 2^4 \cdot 3 \cdot 5 \cdot 7^2 \cdot 13 \cdot 17 \cdot 23 \cdot 47 \cdot 53$$
$$558\,383\,307\,300 = 2^2 \cdot 3^2 \cdot 5^2 \cdot 7^2 \cdot 13 \cdot 17 \cdot 23 \cdot 47 \cdot 53$$
$$1\,889\,912\,732\,400 = 2^4 \cdot 3^2 \cdot 5^2 \cdot 7^2 \cdot 11 \cdot 17 \cdot 23 \cdot 47 \cdot 53$$
$$5\,804\,731\,963\,800 = 2^3 \cdot 3^2 \cdot 5^2 \cdot 7 \cdot 11 \cdot 17 \cdot 23 \cdot 43 \cdot 47 \cdot 53$$
$$16\,253\,249\,498\,640 = 2^4 \cdot 3^2 \cdot 5 \cdot 7^2 \cdot 11 \cdot 17 \cdot 23 \cdot 43 \cdot 47 \cdot 53$$
$$41\,648\,951\,840\,265 = 3^2 \cdot 5 \cdot 7^2 \cdot 11 \cdot 17 \cdot 23 \cdot 41 \cdot 43 \cdot 47 \cdot 53$$
$$97\,997\,533\,741\,800 = 2^3 \cdot 3^2 \cdot 5^2 \cdot 7^2 \cdot 11 \cdot 23 \cdot 41 \cdot 43 \cdot 47 \cdot 53$$
$$212\,327\,989\,773\,900 = 2^2 \cdot 3 \cdot 5^2 \cdot 7^2 \cdot 11 \cdot 13 \cdot 23 \cdot 41 \cdot 43 \cdot 47 \cdot 53$$
$$424\,655\,979\,547\,800 = 2^3 \cdot 3 \cdot 5^2 \cdot 7^2 \cdot 11 \cdot 13 \cdot 23 \cdot 41 \cdot 43 \cdot 47 \cdot 53$$
$$785\,613\,562\,163\,430 = 2 \cdot 3 \cdot 5 \cdot 7^2 \cdot 11 \cdot 13 \cdot 23 \cdot 37 \cdot 41 \cdot 43 \cdot 47 \cdot 53$$
$$1\,346\,766\,106\,565\,880 = 2^3 \cdot 3^2 \cdot 5 \cdot 7 \cdot 11 \cdot 13 \cdot 23 \cdot 37 \cdot 41 \cdot 43 \cdot 47 \cdot 53$$
$$2\,142\,582\,442\,263\,900 = 2^2 \cdot 3^2 \cdot 5^2 \cdot 7^2 \cdot 13 \cdot 23 \cdot 37 \cdot 41 \cdot 43 \cdot 47 \cdot 53$$
$$3\,167\,295\,784\,216\,200 = 2^3 \cdot 3^2 \cdot 5^2 \cdot 7^2 \cdot 13 \cdot 17 \cdot 37 \cdot 41 \cdot 43 \cdot 47 \cdot 53$$
$$4\,355\,031\,703\,297\,275 = 3^2 \cdot 5^2 \cdot 7^2 \cdot 11 \cdot 13 \cdot 17 \cdot 37 \cdot 41 \cdot 43 \cdot 47 \cdot 53$$
$$5\,574\,440\,580\,220\,512 = 2^5 \cdot 3^2 \cdot 7^2 \cdot 11 \cdot 13 \cdot 17 \cdot 37 \cdot 41 \cdot 43 \cdot 47 \cdot 53$$
$$6\,646\,448\,384\,109\,072 = 2^4 \cdot 3^2 \cdot 7^2 \cdot 11 \cdot 17 \cdot 31 \cdot 37 \cdot 41 \cdot 43 \cdot 47 \cdot 53$$
$$7\,384\,942\,649\,010\,080 = 2^5 \cdot 5 \cdot 7^2 \cdot 11 \cdot 17 \cdot 31 \cdot 37 \cdot 41 \cdot 43 \cdot 47 \cdot 53$$
$$7\,648\,690\,600\,760\,440 = 2^3 \cdot 5 \cdot 7 \cdot 11 \cdot 17 \cdot 29 \cdot 31 \cdot 37 \cdot 41 \cdot 43 \cdot 47 \cdot 53$$

<u>Row 57</u>

$$57 = 3 \cdot 19$$
$$1\,596 = 2^2 \cdot 3 \cdot 7 \cdot 19$$
$$29\,260 = 2^2 \cdot 5 \cdot 7 \cdot 11 \cdot 19$$
$$395\,010 = 2 \cdot 3^3 \cdot 5 \cdot 7 \cdot 11 \cdot 19$$
$$4\,187\,106 = 2 \cdot 3^3 \cdot 7 \cdot 11 \cdot 19 \cdot 53$$
$$36\,288\,252 = 2^2 \cdot 3^2 \cdot 7 \cdot 11 \cdot 13 \cdot 19 \cdot 53$$
$$264\,385\,836 = 2^2 \cdot 3^3 \cdot 11 \cdot 13 \cdot 17 \cdot 19 \cdot 53$$
$$1\,652\,411\,475 = 3^3 \cdot 5^2 \cdot 11 \cdot 13 \cdot 17 \cdot 19 \cdot 53$$
$$8\,996\,462\,475 = 3 \cdot 5^2 \cdot 7^2 \cdot 11 \cdot 13 \cdot 17 \cdot 19 \cdot 53$$
$$43\,183\,019\,880 = 2^3 \cdot 3^2 \cdot 5 \cdot 7^2 \cdot 11 \cdot 13 \cdot 17 \cdot 19 \cdot 53$$
$$184\,509\,266\,760 = 2^3 \cdot 3^2 \cdot 5 \cdot 7^2 \cdot 13 \cdot 17 \cdot 19 \cdot 47 \cdot 53$$
$$707\,285\,522\,580 = 2^2 \cdot 3 \cdot 5 \cdot 7^2 \cdot 13 \cdot 17 \cdot 19 \cdot 23 \cdot 47 \cdot 53$$
$$2\,448\,296\,039\,700 = 2^2 \cdot 3^3 \cdot 5^2 \cdot 7^2 \cdot 17 \cdot 19 \cdot 23 \cdot 47 \cdot 53$$
$$7\,694\,644\,696\,200 = 2^3 \cdot 3^3 \cdot 5^2 \cdot 7 \cdot 11 \cdot 17 \cdot 19 \cdot 23 \cdot 47 \cdot 53$$
$$22\,057\,981\,462\,440 = 2^3 \cdot 3^2 \cdot 5 \cdot 7 \cdot 11 \cdot 17 \cdot 19 \cdot 23 \cdot 43 \cdot 47 \cdot 53$$
$$57\,902\,201\,338\,905 = 3^3 \cdot 5 \cdot 7^2 \cdot 11 \cdot 17 \cdot 19 \cdot 23 \cdot 43 \cdot 47 \cdot 53$$
$$139\,646\,485\,582\,065 = 3^3 \cdot 5 \cdot 7^2 \cdot 11 \cdot 19 \cdot 23 \cdot 41 \cdot 43 \cdot 47 \cdot 53$$
$$310\,325\,523\,515\,700 = 2^2 \cdot 3 \cdot 5^2 \cdot 7^2 \cdot 11 \cdot 19 \cdot 23 \cdot 41 \cdot 43 \cdot 47 \cdot 53$$
$$636\,983\,969\,321\,700 = 2^2 \cdot 3^2 \cdot 5^2 \cdot 7^2 \cdot 11 \cdot 13 \cdot 23 \cdot 41 \cdot 43 \cdot 47 \cdot 53$$
$$1\,210\,269\,541\,711\,230 = 2 \cdot 3^2 \cdot 5 \cdot 7^2 \cdot 11 \cdot 13 \cdot 19 \cdot 23 \cdot 41 \cdot 43 \cdot 47 \cdot 53$$
$$2\,132\,379\,668\,729\,310 = 2 \cdot 3 \cdot 5 \cdot 7 \cdot 11 \cdot 13 \cdot 19 \cdot 23 \cdot 37 \cdot 41 \cdot 43 \cdot 47 \cdot 53$$
$$3\,489\,348\,548\,829\,780 = 2^2 \cdot 3^3 \cdot 5 \cdot 7 \cdot 13 \cdot 19 \cdot 23 \cdot 37 \cdot 41 \cdot 43 \cdot 47 \cdot 53$$
$$5\,309\,878\,226\,480\,100 = 2^2 \cdot 3^3 \cdot 5^2 \cdot 7^2 \cdot 13 \cdot 19 \cdot 37 \cdot 41 \cdot 43 \cdot 47 \cdot 53$$
$$7\,522\,327\,487\,513\,475 = 3^2 \cdot 5^2 \cdot 7^2 \cdot 13 \cdot 17 \cdot 19 \cdot 37 \cdot 41 \cdot 43 \cdot 47 \cdot 53$$
$$9\,929\,472\,283\,517\,787 = 3^3 \cdot 7^2 \cdot 11 \cdot 13 \cdot 17 \cdot 19 \cdot 37 \cdot 41 \cdot 43 \cdot 47 \cdot 53$$
$$12\,220\,888\,964\,329\,584 = 2^4 \cdot 3^3 \cdot 7^2 \cdot 11 \cdot 17 \cdot 19 \cdot 37 \cdot 41 \cdot 43 \cdot 47 \cdot 53$$
$$14\,031\,391\,033\,119\,152 = 2^4 \cdot 7^2 \cdot 11 \cdot 17 \cdot 19 \cdot 31 \cdot 37 \cdot 41 \cdot 43 \cdot 47 \cdot 53$$
$$15\,033\,633\,249\,770\,520 = 2^3 \cdot 3 \cdot 5 \cdot 7 \cdot 11 \cdot 17 \cdot 19 \cdot 31 \cdot 37 \cdot 41 \cdot 43 \cdot 47 \cdot 53$$

Row 58

$$58 = 2 \cdot 29$$
$$1\,653 = 3 \cdot 19 \cdot 29$$
$$30\,856 = 2^3 \cdot 7 \cdot 19 \cdot 29$$
$$424\,270 = 2 \cdot 5 \cdot 7 \cdot 11 \cdot 19 \cdot 29$$
$$4\,582\,116 = 2^2 \cdot 3^3 \cdot 7 \cdot 11 \cdot 19 \cdot 29$$
$$40\,475\,358 = 2 \cdot 3^2 \cdot 7 \cdot 11 \cdot 19 \cdot 29 \cdot 53$$
$$300\,674\,088 = 2^3 \cdot 3^2 \cdot 11 \cdot 13 \cdot 19 \cdot 29 \cdot 53$$
$$1\,916\,797\,311 = 3^3 \cdot 11 \cdot 13 \cdot 17 \cdot 19 \cdot 29 \cdot 53$$
$$10\,648\,873\,950 = 2 \cdot 3 \cdot 5^2 \cdot 11 \cdot 13 \cdot 17 \cdot 19 \cdot 29 \cdot 53$$
$$52\,179\,482\,355 = 3 \cdot 5 \cdot 7^2 \cdot 11 \cdot 13 \cdot 17 \cdot 19 \cdot 29 \cdot 53$$
$$227\,692\,286\,640 = 2^4 \cdot 3^2 \cdot 5 \cdot 7^2 \cdot 13 \cdot 17 \cdot 19 \cdot 29 \cdot 53$$
$$891\,794\,789\,340 = 2^2 \cdot 3 \cdot 5 \cdot 7^2 \cdot 13 \cdot 17 \cdot 19 \cdot 29 \cdot 47 \cdot 53$$
$$3\,155\,581\,562\,280 = 2^3 \cdot 3 \cdot 5 \cdot 7^2 \cdot 17 \cdot 19 \cdot 23 \cdot 29 \cdot 47 \cdot 53$$
$$10\,142\,940\,735\,900 = 2^2 \cdot 3^3 \cdot 5^2 \cdot 7 \cdot 17 \cdot 19 \cdot 23 \cdot 29 \cdot 47 \cdot 53$$
$$29\,752\,626\,158\,640 = 2^4 \cdot 3^2 \cdot 5 \cdot 7 \cdot 11 \cdot 17 \cdot 19 \cdot 23 \cdot 29 \cdot 47 \cdot 53$$
$$79\,960\,182\,801\,345 = 3^2 \cdot 5 \cdot 7 \cdot 11 \cdot 17 \cdot 19 \cdot 23 \cdot 29 \cdot 43 \cdot 47 \cdot 53$$
$$197\,548\,686\,920\,970 = 2 \cdot 3^3 \cdot 5 \cdot 7^2 \cdot 11 \cdot 19 \cdot 23 \cdot 29 \cdot 43 \cdot 47 \cdot 53$$
$$449\,972\,009\,097\,765 = 3 \cdot 5 \cdot 7^2 \cdot 11 \cdot 19 \cdot 23 \cdot 29 \cdot 41 \cdot 43 \cdot 47 \cdot 53$$
$$947\,309\,492\,837\,400 = 2^3 \cdot 3 \cdot 5^2 \cdot 7^2 \cdot 11 \cdot 23 \cdot 29 \cdot 41 \cdot 43 \cdot 47 \cdot 53$$
$$1\,847\,253\,511\,032\,930 = 2 \cdot 3^2 \cdot 5 \cdot 7^2 \cdot 11 \cdot 13 \cdot 23 \cdot 29 \cdot 41 \cdot 43 \cdot 47 \cdot 53$$
$$3\,342\,649\,210\,440\,540 = 2^2 \cdot 3 \cdot 5 \cdot 7 \cdot 11 \cdot 13 \cdot 19 \cdot 23 \cdot 29 \cdot 41 \cdot 43 \cdot 47 \cdot 53$$
$$5\,621\,728\,217\,559\,090 = 2 \cdot 3 \cdot 5 \cdot 7 \cdot 13 \cdot 19 \cdot 23 \cdot 29 \cdot 37 \cdot 41 \cdot 43 \cdot 47 \cdot 53$$
$$8\,799\,226\,775\,309\,880 = 2^3 \cdot 3^3 \cdot 5 \cdot 7 \cdot 13 \cdot 19 \cdot 29 \cdot 37 \cdot 41 \cdot 43 \cdot 47 \cdot 53$$
$$12\,832\,205\,713\,993\,575 = 3^2 \cdot 5^2 \cdot 7^2 \cdot 13 \cdot 19 \cdot 29 \cdot 37 \cdot 41 \cdot 43 \cdot 47 \cdot 53$$
$$17\,451\,799\,771\,031\,262 = 2 \cdot 3^2 \cdot 7^2 \cdot 13 \cdot 17 \cdot 19 \cdot 29 \cdot 37 \cdot 41 \cdot 43 \cdot 47 \cdot 53$$
$$22\,150\,361\,247\,847\,371 = 3^3 \cdot 7^2 \cdot 11 \cdot 17 \cdot 19 \cdot 29 \cdot 37 \cdot 41 \cdot 43 \cdot 47 \cdot 53$$
$$26\,252\,279\,997\,448\,736 = 2^5 \cdot 7^2 \cdot 11 \cdot 17 \cdot 19 \cdot 29 \cdot 37 \cdot 41 \cdot 43 \cdot 47 \cdot 53$$
$$29\,065\,024\,282\,889\,672 = 2^3 \cdot 7 \cdot 11 \cdot 17 \cdot 19 \cdot 29 \cdot 31 \cdot 37 \cdot 41 \cdot 43 \cdot 47 \cdot 53$$
$$30\,067\,266\,499\,541\,040 = 2^4 \cdot 3 \cdot 5 \cdot 7 \cdot 11 \cdot 17 \cdot 19 \cdot 31 \cdot 37 \cdot 41 \cdot 43 \cdot 47 \cdot 53$$

Row 59

$$
\begin{aligned}
59 &\ \text{is Prime} \\
1\ 711 &= 29 \cdot 59 \\
32\ 509 &= 19 \cdot 29 \cdot 59 \\
455\ 126 &= 2 \cdot 7 \cdot 19 \cdot 29 \cdot 59 \\
5\ 006\ 386 &= 2 \cdot 7 \cdot 11 \cdot 19 \cdot 29 \cdot 59 \\
45\ 057\ 474 &= 2 \cdot 3^2 \cdot 7 \cdot 11 \cdot 19 \cdot 29 \cdot 59 \\
341\ 149\ 446 &= 2 \cdot 3^2 \cdot 11 \cdot 19 \cdot 29 \cdot 53 \cdot 59 \\
2\ 217\ 471\ 399 &= 3^2 \cdot 11 \cdot 13 \cdot 19 \cdot 29 \cdot 53 \cdot 59 \\
12\ 565\ 671\ 261 &= 3 \cdot 11 \cdot 13 \cdot 17 \cdot 19 \cdot 29 \cdot 53 \cdot 59 \\
62\ 828\ 356\ 305 &= 3 \cdot 5 \cdot 11 \cdot 13 \cdot 17 \cdot 19 \cdot 29 \cdot 53 \cdot 59 \\
279\ 871\ 768\ 995 &= 3 \cdot 5 \cdot 7^2 \cdot 13 \cdot 17 \cdot 19 \cdot 29 \cdot 53 \cdot 59 \\
1\ 119\ 487\ 075\ 980 &= 2^2 \cdot 3 \cdot 5 \cdot 7^2 \cdot 13 \cdot 17 \cdot 19 \cdot 29 \cdot 53 \cdot 59 \\
4\ 047\ 376\ 351\ 620 &= 2^2 \cdot 3 \cdot 5 \cdot 7^2 \cdot 17 \cdot 19 \cdot 29 \cdot 47 \cdot 53 \cdot 59 \\
13\ 298\ 522\ 298\ 180 &= 2^2 \cdot 3 \cdot 5 \cdot 7 \cdot 17 \cdot 19 \cdot 23 \cdot 29 \cdot 47 \cdot 53 \cdot 59 \\
39\ 895\ 566\ 894\ 540 &= 2^2 \cdot 3^2 \cdot 5 \cdot 7 \cdot 17 \cdot 19 \cdot 23 \cdot 29 \cdot 47 \cdot 53 \cdot 59 \\
109\ 712\ 808\ 959\ 985 &= 3^2 \cdot 5 \cdot 7 \cdot 11 \cdot 17 \cdot 19 \cdot 23 \cdot 29 \cdot 47 \cdot 53 \cdot 59 \\
277\ 508\ 869\ 722\ 315 &= 3^2 \cdot 5 \cdot 7 \cdot 11 \cdot 19 \cdot 23 \cdot 29 \cdot 43 \cdot 47 \cdot 53 \cdot 59 \\
647\ 520\ 696\ 018\ 735 &= 3 \cdot 5 \cdot 7^2 \cdot 11 \cdot 19 \cdot 23 \cdot 29 \cdot 43 \cdot 47 \cdot 53 \cdot 59 \\
1\ 397\ 281\ 501\ 935\ 165 &= 3 \cdot 5 \cdot 7^2 \cdot 11 \cdot 23 \cdot 29 \cdot 41 \cdot 43 \cdot 47 \cdot 53 \cdot 59 \\
2\ 794\ 563\ 003\ 870\ 330 &= 2 \cdot 3 \cdot 5 \cdot 7^2 \cdot 11 \cdot 23 \cdot 29 \cdot 41 \cdot 43 \cdot 47 \cdot 53 \cdot 59 \\
5\ 189\ 902\ 721\ 473\ 470 &= 2 \cdot 3 \cdot 5 \cdot 7 \cdot 11 \cdot 13 \cdot 23 \cdot 29 \cdot 41 \cdot 43 \cdot 47 \cdot 53 \cdot 59 \\
8\ 964\ 377\ 427\ 999\ 630 &= 2 \cdot 3 \cdot 5 \cdot 7 \cdot 13 \cdot 19 \cdot 23 \cdot 29 \cdot 41 \cdot 43 \cdot 47 \cdot 53 \cdot 59 \\
14\ 420\ 954\ 992\ 868\ 970 &= 2 \cdot 3 \cdot 5 \cdot 7 \cdot 13 \cdot 19 \cdot 29 \cdot 37 \cdot 41 \cdot 43 \cdot 47 \cdot 53 \cdot 59 \\
21\ 631\ 432\ 489\ 303\ 455 &= 3^2 \cdot 5 \cdot 7 \cdot 13 \cdot 19 \cdot 29 \cdot 37 \cdot 41 \cdot 43 \cdot 47 \cdot 53 \cdot 59 \\
30\ 284\ 005\ 485\ 024\ 837 &= 3^2 \cdot 7^2 \cdot 13 \cdot 19 \cdot 29 \cdot 37 \cdot 41 \cdot 43 \cdot 47 \cdot 53 \cdot 59 \\
39\ 602\ 161\ 018\ 878\ 633 &= 3^2 \cdot 7^2 \cdot 17 \cdot 19 \cdot 29 \cdot 37 \cdot 41 \cdot 43 \cdot 47 \cdot 53 \cdot 59 \\
48\ 402\ 641\ 245\ 296\ 107 &= 7^2 \cdot 11 \cdot 17 \cdot 19 \cdot 29 \cdot 37 \cdot 41 \cdot 43 \cdot 47 \cdot 53 \cdot 59 \\
55\ 317\ 304\ 280\ 338\ 408 &= 2^3 \cdot 7 \cdot 11 \cdot 17 \cdot 19 \cdot 29 \cdot 37 \cdot 41 \cdot 43 \cdot 47 \cdot 53 \cdot 59 \\
59\ 132\ 290\ 782\ 430\ 712 &= 2^3 \cdot 7 \cdot 11 \cdot 17 \cdot 19 \cdot 31 \cdot 37 \cdot 41 \cdot 43 \cdot 47 \cdot 53 \cdot 59
\end{aligned}
$$

Pascal's Triangle — Prime Factorization — To Center Number (omitting 1's)

Row 60

$$
\begin{aligned}
60 &= 2^2 \cdot 3 \cdot 5 \\
1\,770 &= 2 \cdot 3 \cdot 5 \cdot 59 \\
34\,220 &= 2^2 \cdot 5 \cdot 29 \cdot 59 \\
487\,635 &= 3 \cdot 5 \cdot 19 \cdot 29 \cdot 59 \\
5\,461\,512 &= 2^3 \cdot 3 \cdot 7 \cdot 19 \cdot 29 \cdot 59 \\
50\,063\,860 &= 2^2 \cdot 5 \cdot 7 \cdot 11 \cdot 19 \cdot 29 \cdot 59 \\
000\,200\,920 &= 2^3 \cdot 0^3 \cdot 5 \cdot 11 \cdot 19 \cdot 29 \cdot 59 \\
2\,558\,620\,845 &= 3^3 \cdot 5 \cdot 11 \cdot 19 \cdot 29 \cdot 53 \cdot 59 \\
14\,783\,142\,660 &= 2^2 \cdot 3 \cdot 5 \cdot 11 \cdot 13 \cdot 19 \cdot 29 \cdot 53 \cdot 59 \\
75\,394\,027\,566 &= 2 \cdot 3^2 \cdot 11 \cdot 13 \cdot 17 \cdot 19 \cdot 29 \cdot 53 \cdot 59 \\
342\,700\,125\,300 &= 2^2 \cdot 3^2 \cdot 5^2 \cdot 13 \cdot 17 \cdot 19 \cdot 29 \cdot 53 \cdot 59 \\
1\,399\,358\,844\,975 &= 3 \cdot 5^2 \cdot 7^2 \cdot 13 \cdot 17 \cdot 19 \cdot 29 \cdot 53 \cdot 59 \\
5\,166\,863\,427\,600 &= 2^4 \cdot 3^2 \cdot 5^2 \cdot 7^2 \cdot 17 \cdot 19 \cdot 29 \cdot 53 \cdot 59 \\
17\,345\,898\,649\,800 &= 2^3 \cdot 3^2 \cdot 5^2 \cdot 7 \cdot 17 \cdot 19 \cdot 29 \cdot 47 \cdot 53 \cdot 59 \\
53\,194\,089\,192\,720 &= 2^4 \cdot 3 \cdot 5 \cdot 7 \cdot 17 \cdot 19 \cdot 23 \cdot 29 \cdot 47 \cdot 53 \cdot 59 \\
149\,608\,375\,854\,525 &= 3^3 \cdot 5^2 \cdot 7 \cdot 17 \cdot 19 \cdot 23 \cdot 29 \cdot 47 \cdot 53 \cdot 59 \\
387\,221\,678\,682\,300 &= 2^2 \cdot 3^3 \cdot 5^2 \cdot 7 \cdot 11 \cdot 19 \cdot 23 \cdot 29 \cdot 47 \cdot 53 \cdot 59 \\
925\,029\,565\,741\,050 &= 2 \cdot 3 \cdot 5^2 \cdot 7 \cdot 11 \cdot 19 \cdot 23 \cdot 29 \cdot 43 \cdot 47 \cdot 53 \cdot 59 \\
2\,044\,802\,197\,953\,900 &= 2^2 \cdot 3^2 \cdot 5^2 \cdot 7^2 \cdot 11 \cdot 23 \cdot 29 \cdot 43 \cdot 47 \cdot 53 \cdot 59 \\
4\,191\,844\,505\,805\,495 &= 3^2 \cdot 5 \cdot 7^2 \cdot 11 \cdot 23 \cdot 29 \cdot 41 \cdot 43 \cdot 47 \cdot 53 \cdot 59 \\
7\,984\,465\,725\,343\,800 &= 2^3 \cdot 3 \cdot 5^2 \cdot 7 \cdot 11 \cdot 23 \cdot 29 \cdot 41 \cdot 43 \cdot 47 \cdot 53 \cdot 59 \\
14\,154\,280\,149\,473\,100 &= 2^2 \cdot 3^2 \cdot 5^2 \cdot 7 \cdot 13 \cdot 23 \cdot 29 \cdot 41 \cdot 43 \cdot 47 \cdot 53 \cdot 59 \\
23\,385\,332\,420\,868\,600 &= 2^3 \cdot 3^2 \cdot 5^2 \cdot 7 \cdot 13 \cdot 19 \cdot 29 \cdot 41 \cdot 43 \cdot 47 \cdot 53 \cdot 59 \\
36\,052\,387\,482\,172\,425 &= 3 \cdot 5^2 \cdot 7 \cdot 13 \cdot 19 \cdot 29 \cdot 37 \cdot 41 \cdot 43 \cdot 47 \cdot 53 \cdot 59 \\
51\,915\,437\,974\,328\,292 &= 2^2 \cdot 3^3 \cdot 7 \cdot 13 \cdot 19 \cdot 29 \cdot 37 \cdot 41 \cdot 43 \cdot 47 \cdot 53 \cdot 59 \\
69\,886\,166\,503\,903\,470 &= 2 \cdot 3^3 \cdot 5 \cdot 7^2 \cdot 19 \cdot 29 \cdot 37 \cdot 41 \cdot 43 \cdot 47 \cdot 53 \cdot 59 \\
88\,004\,802\,264\,174\,740 &= 2^2 \cdot 5 \cdot 7^2 \cdot 17 \cdot 19 \cdot 29 \cdot 37 \cdot 41 \cdot 43 \cdot 47 \cdot 53 \cdot 59 \\
103\,719\,945\,525\,634\,515 &= 3 \cdot 5 \cdot 7 \cdot 11 \cdot 17 \cdot 19 \cdot 29 \cdot 37 \cdot 41 \cdot 43 \cdot 47 \cdot 53 \cdot 59 \\
114\,449\,595\,062\,769\,120 &= 2^5 \cdot 3 \cdot 5 \cdot 7 \cdot 11 \cdot 17 \cdot 19 \cdot 37 \cdot 41 \cdot 43 \cdot 47 \cdot 53 \cdot 59 \\
118\,264\,581\,564\,861\,424 &= 2^4 \cdot 7 \cdot 11 \cdot 17 \cdot 19 \cdot 31 \cdot 37 \cdot 41 \cdot 43 \cdot 47 \cdot 53 \cdot 59
\end{aligned}
$$

Row 61

$$
\begin{aligned}
61 &\ \text{is Prime} \\
1\,830 &= 2 \cdot 3 \cdot 5 \cdot 61 \\
35\,990 &= 2 \cdot 5 \cdot 59 \cdot 61 \\
521\,855 &= 5 \cdot 29 \cdot 59 \cdot 61 \\
5\,949\,147 &= 3 \cdot 19 \cdot 29 \cdot 59 \cdot 61 \\
55\,525\,372 &= 2^2 \cdot 7 \cdot 19 \cdot 29 \cdot 59 \cdot 61 \\
436\,270\,780 &= 2^2 \cdot 5 \cdot 11 \cdot 19 \cdot 29 \cdot 59 \cdot 61 \\
2\,944\,827\,765 &= 3^3 \cdot 5 \cdot 11 \cdot 19 \cdot 29 \cdot 59 \cdot 61 \\
17\,341\,763\,505 &= 3 \cdot 5 \cdot 11 \cdot 19 \cdot 29 \cdot 53 \cdot 59 \cdot 61 \\
90\,177\,170\,226 &= 2 \cdot 3 \cdot 11 \cdot 13 \cdot 19 \cdot 29 \cdot 53 \cdot 59 \cdot 61 \\
418\,094\,152\,866 &= 2 \cdot 3^2 \cdot 13 \cdot 17 \cdot 19 \cdot 29 \cdot 53 \cdot 59 \cdot 61 \\
1\,742\,058\,970\,275 &= 3 \cdot 5^2 \cdot 13 \cdot 17 \cdot 19 \cdot 29 \cdot 53 \cdot 59 \cdot 61 \\
6\,566\,222\,272\,575 &= 3 \cdot 5^2 \cdot 7^2 \cdot 17 \cdot 19 \cdot 29 \cdot 53 \cdot 59 \cdot 61 \\
22\,512\,762\,077\,400 &= 2^3 \cdot 3^2 \cdot 5^2 \cdot 7 \cdot 17 \cdot 19 \cdot 29 \cdot 53 \cdot 59 \cdot 61 \\
70\,539\,987\,842\,520 &= 2^3 \cdot 3 \cdot 5 \cdot 7 \cdot 17 \cdot 19 \cdot 29 \cdot 47 \cdot 53 \cdot 59 \cdot 61 \\
202\,802\,465\,047\,245 &= 3 \cdot 5 \cdot 7 \cdot 17 \cdot 19 \cdot 23 \cdot 29 \cdot 47 \cdot 53 \cdot 59 \cdot 61 \\
536\,830\,054\,536\,825 &= 3^3 \cdot 5^2 \cdot 7 \cdot 19 \cdot 23 \cdot 29 \cdot 47 \cdot 53 \cdot 59 \cdot 61 \\
1\,312\,251\,244\,423\,350 &= 2 \cdot 3 \cdot 5^2 \cdot 7 \cdot 11 \cdot 19 \cdot 23 \cdot 29 \cdot 47 \cdot 53 \cdot 59 \cdot 61 \\
2\,969\,831\,763\,694\,950 &= 2 \cdot 3 \cdot 5^2 \cdot 7 \cdot 11 \cdot 23 \cdot 29 \cdot 43 \cdot 47 \cdot 53 \cdot 59 \cdot 61 \\
6\,236\,646\,703\,759\,395 &= 3^2 \cdot 5 \cdot 7^2 \cdot 11 \cdot 23 \cdot 29 \cdot 43 \cdot 47 \cdot 53 \cdot 59 \cdot 61 \\
12\,176\,310\,231\,149\,295 &= 3 \cdot 5 \cdot 7 \cdot 11 \cdot 23 \cdot 29 \cdot 41 \cdot 43 \cdot 47 \cdot 53 \cdot 59 \cdot 61 \\
22\,138\,745\,874\,816\,900 &= 2^2 \cdot 3 \cdot 5^2 \cdot 7 \cdot 23 \cdot 29 \cdot 41 \cdot 43 \cdot 47 \cdot 53 \cdot 59 \cdot 61 \\
37\,539\,612\,570\,341\,700 &= 2^2 \cdot 3^2 \cdot 5^2 \cdot 7 \cdot 13 \cdot 29 \cdot 41 \cdot 43 \cdot 47 \cdot 53 \cdot 59 \cdot 61 \\
59\,437\,719\,903\,041\,025 &= 3 \cdot 5^2 \cdot 7 \cdot 13 \cdot 19 \cdot 29 \cdot 41 \cdot 43 \cdot 47 \cdot 53 \cdot 59 \cdot 61 \\
87\,967\,825\,456\,500\,717 &= 3 \cdot 7 \cdot 13 \cdot 19 \cdot 29 \cdot 37 \cdot 41 \cdot 43 \cdot 47 \cdot 53 \cdot 59 \cdot 61 \\
121\,801\,604\,478\,231\,762 &= 2 \cdot 3^3 \cdot 7 \cdot 19 \cdot 29 \cdot 37 \cdot 41 \cdot 43 \cdot 47 \cdot 53 \cdot 59 \cdot 61 \\
157\,890\,968\,768\,078\,210 &= 2 \cdot 5 \cdot 7^2 \cdot 19 \cdot 29 \cdot 37 \cdot 41 \cdot 43 \cdot 47 \cdot 53 \cdot 59 \cdot 61 \\
191\,724\,747\,789\,809\,255 &= 5 \cdot 7 \cdot 17 \cdot 19 \cdot 29 \cdot 37 \cdot 41 \cdot 43 \cdot 47 \cdot 53 \cdot 59 \cdot 61 \\
218\,169\,540\,588\,403\,635 &= 3 \cdot 5 \cdot 7 \cdot 11 \cdot 17 \cdot 19 \cdot 37 \cdot 41 \cdot 43 \cdot 47 \cdot 53 \cdot 59 \cdot 61 \\
232\,714\,176\,627\,630\,544 &= 2^4 \cdot 7 \cdot 11 \cdot 17 \cdot 19 \cdot 37 \cdot 41 \cdot 43 \cdot 47 \cdot 53 \cdot 59 \cdot 61
\end{aligned}
$$

Row 62

$$62 = 2 \cdot 31$$

$$1\,891 = 31 \cdot 61$$

$$37\,820 = 2^2 \cdot 5 \cdot 31 \cdot 61$$

$$557\,845 = 5 \cdot 31 \cdot 59 \cdot 61$$

$$6\,471\,002 = 2 \cdot 29 \cdot 31 \cdot 59 \cdot 61$$

$$61\,474\,519 = 19 \cdot 29 \cdot 31 \cdot 59 \cdot 61$$

$$401\,790\,152 = 2^3 \cdot 19 \cdot 29 \cdot 31 \cdot 59 \cdot 61$$

$$3\,381\,098\,545 = 5 \cdot 11 \cdot 19 \cdot 29 \cdot 31 \cdot 59 \cdot 61$$

$$20\,286\,591\,270 = 2 \cdot 3 \cdot 5 \cdot 11 \cdot 19 \cdot 29 \cdot 31 \cdot 59 \cdot 61$$

$$107\,518\,933\,731 = 3 \cdot 11 \cdot 19 \cdot 29 \cdot 31 \cdot 53 \cdot 59 \cdot 61$$

$$508\,271\,323\,092 = 2^2 \cdot 3 \cdot 13 \cdot 19 \cdot 29 \cdot 31 \cdot 53 \cdot 59 \cdot 61$$

$$2\,160\,153\,123\,141 = 3 \cdot 13 \cdot 17 \cdot 19 \cdot 29 \cdot 31 \cdot 53 \cdot 59 \cdot 61$$

$$8\,308\,281\,242\,850 = 2 \cdot 3 \cdot 5^2 \cdot 17 \cdot 19 \cdot 29 \cdot 31 \cdot 53 \cdot 59 \cdot 61$$

$$29\,078\,984\,349\,975 = 3 \cdot 5^2 \cdot 7 \cdot 17 \cdot 19 \cdot 29 \cdot 31 \cdot 53 \cdot 59 \cdot 61$$

$$93\,052\,749\,919\,920 = 2^4 \cdot 3 \cdot 5 \cdot 7 \cdot 17 \cdot 19 \cdot 29 \cdot 31 \cdot 53 \cdot 59 \cdot 61$$

$$273\,342\,452\,889\,765 = 3 \cdot 5 \cdot 7 \cdot 17 \cdot 19 \cdot 29 \cdot 31 \cdot 47 \cdot 53 \cdot 59 \cdot 61$$

$$739\,632\,519\,584\,070 = 2 \cdot 3 \cdot 5 \cdot 7 \cdot 19 \cdot 23 \cdot 29 \cdot 31 \cdot 47 \cdot 53 \cdot 59 \cdot 61$$

$$1\,849\,081\,298\,960\,175 = 3 \cdot 5^2 \cdot 7 \cdot 19 \cdot 23 \cdot 29 \cdot 31 \cdot 47 \cdot 53 \cdot 59 \cdot 61$$

$$4\,282\,083\,008\,118\,300 = 2^2 \cdot 3 \cdot 5^2 \cdot 7 \cdot 11 \cdot 23 \cdot 29 \cdot 31 \cdot 47 \cdot 53 \cdot 59 \cdot 61$$

$$9\,206\,478\,467\,454\,345 = 3 \cdot 5 \cdot 7 \cdot 11 \cdot 23 \cdot 29 \cdot 31 \cdot 43 \cdot 47 \cdot 53 \cdot 59 \cdot 61$$

$$18\,412\,956\,934\,908\,690 = 2 \cdot 3 \cdot 5 \cdot 7 \cdot 11 \cdot 23 \cdot 29 \cdot 31 \cdot 43 \cdot 47 \cdot 53 \cdot 59 \cdot 61$$

$$34\,315\,056\,105\,966\,195 = 3 \cdot 5 \cdot 7 \cdot 23 \cdot 29 \cdot 31 \cdot 41 \cdot 43 \cdot 47 \cdot 53 \cdot 59 \cdot 61$$

$$59\,678\,358\,445\,158\,600 = 2^3 \cdot 3 \cdot 5^2 \cdot 7 \cdot 29 \cdot 31 \cdot 41 \cdot 43 \cdot 47 \cdot 53 \cdot 59 \cdot 61$$

$$96\,977\,332\,473\,382\,725 = 3 \cdot 5^2 \cdot 7 \cdot 13 \cdot 29 \cdot 31 \cdot 41 \cdot 43 \cdot 47 \cdot 53 \cdot 59 \cdot 61$$

$$147\,405\,545\,359\,541\,742 = 2 \cdot 3 \cdot 7 \cdot 13 \cdot 19 \cdot 29 \cdot 31 \cdot 41 \cdot 43 \cdot 47 \cdot 53 \cdot 59 \cdot 61$$

$$209\,769\,429\,934\,732\,479 = 3 \cdot 7 \cdot 19 \cdot 29 \cdot 31 \cdot 37 \cdot 41 \cdot 43 \cdot 47 \cdot 53 \cdot 59 \cdot 61$$

$$279\,692\,573\,246\,309\,972 = 2^2 \cdot 7 \cdot 19 \cdot 29 \cdot 31 \cdot 37 \cdot 41 \cdot 43 \cdot 47 \cdot 53 \cdot 59 \cdot 61$$

$$349\,615\,716\,557\,887\,465 = 5 \cdot 7 \cdot 19 \cdot 29 \cdot 31 \cdot 37 \cdot 41 \cdot 43 \cdot 47 \cdot 53 \cdot 59 \cdot 61$$

$$409\,894\,288\,378\,212\,890 = 2 \cdot 5 \cdot 7 \cdot 17 \cdot 19 \cdot 31 \cdot 37 \cdot 41 \cdot 43 \cdot 47 \cdot 53 \cdot 59 \cdot 61$$

$$450\,883\,717\,216\,034\,179 = 7 \cdot 11 \cdot 17 \cdot 19 \cdot 31 \cdot 37 \cdot 41 \cdot 43 \cdot 47 \cdot 53 \cdot 59 \cdot 61$$

$$465\,428\,353\,255\,261\,088 = 2^5 \cdot 7 \cdot 11 \cdot 17 \cdot 19 \cdot 37 \cdot 41 \cdot 43 \cdot 47 \cdot 53 \cdot 59 \cdot 61$$

<u>Row 63</u>

$$63 = 3^2 \cdot 7$$
$$1\,953 = 3^2 \cdot 7 \cdot 31$$
$$39\,711 = 3 \cdot 7 \cdot 31 \cdot 61$$
$$595\,665 = 3^2 \cdot 5 \cdot 7 \cdot 31 \cdot 61$$
$$7\,028\,847 = 3^2 \cdot 7 \cdot 31 \cdot 59 \cdot 61$$
$$67\,945\,521 = 3 \cdot 7 \cdot 29 \cdot 31 \cdot 59 \cdot 61$$
$$553\,270\,671 = 3^2 \cdot 19 \cdot 29 \cdot 31 \cdot 59 \cdot 61$$
$$3\,872\,894\,697 = 3^2 \cdot 7 \cdot 19 \cdot 29 \cdot 31 \cdot 59 \cdot 61$$
$$23\,667\,689\,815 = 5 \cdot 7 \cdot 11 \cdot 19 \cdot 29 \cdot 31 \cdot 59 \cdot 61$$
$$127\,805\,525\,001 = 3^3 \cdot 7 \cdot 11 \cdot 19 \cdot 29 \cdot 31 \cdot 59 \cdot 61$$
$$615\,790\,256\,823 = 3^3 \cdot 7 \cdot 19 \cdot 29 \cdot 31 \cdot 53 \cdot 59 \cdot 61$$
$$2\,668\,424\,446\,233 = 3^2 \cdot 7 \cdot 13 \cdot 19 \cdot 29 \cdot 31 \cdot 53 \cdot 59 \cdot 61$$
$$10\,468\,434\,365\,991 = 3^3 \cdot 7 \cdot 17 \cdot 19 \cdot 29 \cdot 31 \cdot 53 \cdot 59 \cdot 61$$
$$37\,387\,265\,592\,825 = 3^3 \cdot 5^2 \cdot 17 \cdot 19 \cdot 29 \cdot 31 \cdot 53 \cdot 59 \cdot 61$$
$$122\,131\,734\,269\,895 = 3^2 \cdot 5 \cdot 7^2 \cdot 17 \cdot 19 \cdot 29 \cdot 31 \cdot 53 \cdot 59 \cdot 61$$
$$366\,395\,202\,809\,685 = 3^3 \cdot 5 \cdot 7^2 \cdot 17 \cdot 19 \cdot 29 \cdot 31 \cdot 53 \cdot 59 \cdot 61$$
$$1\,012\,974\,972\,473\,835 = 3^3 \cdot 5 \cdot 7^2 \cdot 19 \cdot 29 \cdot 31 \cdot 47 \cdot 53 \cdot 59 \cdot 61$$
$$2\,588\,713\,818\,544\,245 = 3 \cdot 5 \cdot 7^2 \cdot 19 \cdot 23 \cdot 29 \cdot 31 \cdot 47 \cdot 53 \cdot 59 \cdot 61$$
$$6\,131\,164\,307\,078\,475 = 3^3 \cdot 5^2 \cdot 7^2 \cdot 23 \cdot 29 \cdot 31 \cdot 47 \cdot 53 \cdot 59 \cdot 61$$
$$13\,488\,561\,475\,572\,645 = 3^3 \cdot 5 \cdot 7^2 \cdot 11 \cdot 23 \cdot 29 \cdot 31 \cdot 47 \cdot 53 \cdot 59 \cdot 61$$
$$27\,619\,435\,402\,363\,035 = 3^2 \cdot 5 \cdot 7 \cdot 11 \cdot 23 \cdot 29 \cdot 31 \cdot 43 \cdot 47 \cdot 53 \cdot 59 \cdot 61$$
$$52\,728\,013\,040\,874\,885 = 3^3 \cdot 5 \cdot 7^2 \cdot 23 \cdot 29 \cdot 31 \cdot 43 \cdot 47 \cdot 53 \cdot 59 \cdot 61$$
$$93\,993\,414\,551\,124\,795 = 3^3 \cdot 5 \cdot 7^2 \cdot 29 \cdot 31 \cdot 41 \cdot 43 \cdot 47 \cdot 53 \cdot 59 \cdot 61$$
$$156\,655\,690\,918\,541\,325 = 3^2 \cdot 5^2 \cdot 7^2 \cdot 29 \cdot 31 \cdot 41 \cdot 43 \cdot 47 \cdot 53 \cdot 59 \cdot 61$$
$$244\,382\,877\,832\,924\,467 = 3^3 \cdot 7^2 \cdot 13 \cdot 29 \cdot 31 \cdot 41 \cdot 43 \cdot 47 \cdot 53 \cdot 59 \cdot 61$$
$$357\,174\,975\,294\,274\,221 = 3^3 \cdot 7^2 \cdot 19 \cdot 29 \cdot 31 \cdot 41 \cdot 43 \cdot 47 \cdot 53 \cdot 59 \cdot 61$$
$$489\,462\,003\,181\,042\,451 = 7^2 \cdot 19 \cdot 29 \cdot 31 \cdot 37 \cdot 41 \cdot 43 \cdot 47 \cdot 53 \cdot 59 \cdot 61$$
$$629\,308\,289\,804\,197\,437 = 3^2 \cdot 7 \cdot 19 \cdot 29 \cdot 31 \cdot 37 \cdot 41 \cdot 43 \cdot 47 \cdot 53 \cdot 59 \cdot 61$$
$$759\,510\,004\,936\,100\,355 = 3^2 \cdot 5 \cdot 7^2 \cdot 19 \cdot 31 \cdot 37 \cdot 41 \cdot 43 \cdot 47 \cdot 53 \cdot 59 \cdot 61$$
$$860\,778\,005\,594\,247\,069 = 3 \cdot 7^2 \cdot 17 \cdot 19 \cdot 31 \cdot 37 \cdot 41 \cdot 43 \cdot 47 \cdot 53 \cdot 59 \cdot 61$$
$$916\,312\,070\,471\,295\,267 = 3^2 \cdot 7^2 \cdot 11 \cdot 17 \cdot 19 \cdot 37 \cdot 41 \cdot 43 \cdot 47 \cdot 53 \cdot 59 \cdot 61$$

Row 64

$$64 = 2^6$$

$$2\,016 = 2^5 \cdot 3^2 \cdot 7$$

$$41\,664 = 2^6 \cdot 3 \cdot 7 \cdot 31$$

$$635\,376 = 2^4 \cdot 3 \cdot 7 \cdot 31 \cdot 61$$

$$7\,624\,512 = 2^6 \cdot 3^2 \cdot 7 \cdot 31 \cdot 61$$

$$74\,974\,368 = 2^5 \cdot 3 \cdot 7 \cdot 31 \cdot 59 \cdot 61$$

$$621\,216\,192 = 2^6 \cdot 3 \cdot 29 \cdot 31 \cdot 59 \cdot 61$$

$$4\,426\,165\,368 = 2^3 \cdot 3^2 \cdot 19 \cdot 29 \cdot 31 \cdot 59 \cdot 61$$

$$27\,540\,584\,512 = 2^6 \cdot 7 \cdot 19 \cdot 29 \cdot 31 \cdot 59 \cdot 61$$

$$151\,473\,214\,816 = 2^5 \cdot 7 \cdot 11 \cdot 19 \cdot 29 \cdot 31 \cdot 59 \cdot 61$$

$$743\,595\,781\,824 = 2^6 \cdot 3^3 \cdot 7 \cdot 19 \cdot 29 \cdot 31 \cdot 59 \cdot 61$$

$$3\,284\,214\,703\,056 = 2^4 \cdot 3^2 \cdot 7 \cdot 19 \cdot 29 \cdot 31 \cdot 53 \cdot 59 \cdot 61$$

$$13\,136\,858\,812\,224 = 2^6 \cdot 3^2 \cdot 7 \cdot 19 \cdot 29 \cdot 31 \cdot 53 \cdot 59 \cdot 61$$

$$47\,855\,699\,958\,816 = 2^5 \cdot 3^3 \cdot 17 \cdot 19 \cdot 29 \cdot 31 \cdot 53 \cdot 59 \cdot 61$$

$$159\,518\,999\,862\,720 = 2^6 \cdot 3^2 \cdot 5 \cdot 17 \cdot 19 \cdot 29 \cdot 31 \cdot 53 \cdot 59 \cdot 61$$

$$488\,526\,937\,079\,580 = 2^2 \cdot 3^2 \cdot 5 \cdot 7^2 \cdot 17 \cdot 19 \cdot 29 \cdot 31 \cdot 53 \cdot 59 \cdot 61$$

$$1\,379\,370\,175\,283\,520 = 2^6 \cdot 3^3 \cdot 5 \cdot 7^2 \cdot 19 \cdot 29 \cdot 31 \cdot 53 \cdot 59 \cdot 61$$

$$3\,601\,688\,791\,018\,080 = 2^5 \cdot 3 \cdot 5 \cdot 7^2 \cdot 19 \cdot 29 \cdot 31 \cdot 47 \cdot 53 \cdot 59 \cdot 61$$

$$8\,719\,878\,125\,622\,720 = 2^6 \cdot 3 \cdot 5 \cdot 7^2 \cdot 23 \cdot 29 \cdot 31 \cdot 47 \cdot 53 \cdot 59 \cdot 61$$

$$19\,619\,725\,782\,651\,120 = 2^4 \cdot 3^3 \cdot 5 \cdot 7^2 \cdot 23 \cdot 29 \cdot 31 \cdot 47 \cdot 53 \cdot 59 \cdot 61$$

$$41\,107\,996\,877\,935\,680 = 2^6 \cdot 3^2 \cdot 5 \cdot 7 \cdot 11 \cdot 23 \cdot 29 \cdot 31 \cdot 47 \cdot 53 \cdot 59 \cdot 61$$

$$80\,347\,448\,443\,237\,920 = 2^5 \cdot 3^2 \cdot 5 \cdot 7 \cdot 23 \cdot 29 \cdot 31 \cdot 43 \cdot 47 \cdot 53 \cdot 59 \cdot 61$$

$$146\,721\,427\,591\,999\,680 = 2^6 \cdot 3^3 \cdot 5 \cdot 7^2 \cdot 29 \cdot 31 \cdot 43 \cdot 47 \cdot 53 \cdot 59 \cdot 61$$

$$250\,649\,105\,469\,666\,120 = 2^3 \cdot 3^2 \cdot 5 \cdot 7^2 \cdot 29 \cdot 31 \cdot 41 \cdot 43 \cdot 47 \cdot 53 \cdot 59 \cdot 61$$

$$401\,038\,568\,751\,465\,792 = 2^6 \cdot 3^2 \cdot 7^2 \cdot 29 \cdot 31 \cdot 41 \cdot 43 \cdot 47 \cdot 53 \cdot 59 \cdot 61$$

$$601\,557\,853\,127\,198\,688 = 2^5 \cdot 3^3 \cdot 7^2 \cdot 29 \cdot 31 \cdot 41 \cdot 43 \cdot 47 \cdot 53 \cdot 59 \cdot 61$$

$$846\,636\,978\,475\,316\,672 = 2^6 \cdot 7^2 \cdot 19 \cdot 29 \cdot 31 \cdot 41 \cdot 43 \cdot 47 \cdot 53 \cdot 59 \cdot 61$$

$$1\,118\,770\,292\,985\,239\,888 = 2^4 \cdot 7 \cdot 19 \cdot 29 \cdot 31 \cdot 37 \cdot 41 \cdot 43 \cdot 47 \cdot 53 \cdot 59 \cdot 61$$

$$1\,388\,818\,294\,740\,297\,792 = 2^6 \cdot 3^2 \cdot 7 \cdot 19 \cdot 31 \cdot 37 \cdot 41 \cdot 43 \cdot 47 \cdot 53 \cdot 59 \cdot 61$$

$$1\,620\,288\,010\,530\,347\,424 = 2^5 \cdot 3 \cdot 7^2 \cdot 19 \cdot 31 \cdot 37 \cdot 41 \cdot 43 \cdot 47 \cdot 53 \cdot 59 \cdot 61$$

$$1\,777\,090\,076\,065\,542\,336 = 2^6 \cdot 3 \cdot 7^2 \cdot 17 \cdot 19 \cdot 37 \cdot 41 \cdot 43 \cdot 47 \cdot 53 \cdot 59 \cdot 61$$

$$1\,832\,624\,140\,942\,590\,534 = 2 \cdot 3^2 \cdot 7^2 \cdot 11 \cdot 17 \cdot 19 \cdot 37 \cdot 41 \cdot 43 \cdot 47 \cdot 53 \cdot 59 \cdot 61$$

<u>Row 65</u>

$65 = 5 \cdot 13$

$2\,080 = 2^5 \cdot 5 \cdot 13$

$43\,680 = 2^5 \cdot 3 \cdot 5 \cdot 7 \cdot 13$

$677\,040 = 2^4 \cdot 3 \cdot 5 \cdot 7 \cdot 13 \cdot 31$

$8\,259\,888 = 2^4 \cdot 3 \cdot 7 \cdot 13 \cdot 31 \cdot 61$

$82\,598\,880 = 2^5 \cdot 3 \cdot 5 \cdot 7 \cdot 13 \cdot 31 \cdot 61$

$696\,190\,560 = 2^5 \cdot 3 \cdot 5 \cdot 13 \cdot 31 \cdot 59 \cdot 61$

$5\,047\,381\,560 = 2^3 \cdot 3 \cdot 5 \cdot 13 \cdot 29 \cdot 31 \cdot 59 \cdot 61$

$31\,966\,749\,880 = 2^3 \cdot 5 \cdot 13 \cdot 19 \cdot 29 \cdot 31 \cdot 59 \cdot 61$

$179\,013\,799\,328 = 2^5 \cdot 7 \cdot 13 \cdot 19 \cdot 29 \cdot 31 \cdot 59 \cdot 61$

$895\,068\,996\,640 = 2^5 \cdot 5 \cdot 7 \cdot 13 \cdot 19 \cdot 29 \cdot 31 \cdot 59 \cdot 61$

$4\,027\,810\,484\,880 = 2^4 \cdot 3^2 \cdot 5 \cdot 7 \cdot 13 \cdot 19 \cdot 29 \cdot 31 \cdot 59 \cdot 61$

$16\,421\,073\,515\,280 = 2^4 \cdot 3^2 \cdot 5 \cdot 7 \cdot 19 \cdot 29 \cdot 31 \cdot 53 \cdot 59 \cdot 61$

$60\,992\,558\,771\,040 = 2^5 \cdot 3^2 \cdot 5 \cdot 13 \cdot 19 \cdot 29 \cdot 31 \cdot 53 \cdot 59 \cdot 61$

$207\,374\,699\,821\,536 = 2^5 \cdot 3^2 \cdot 13 \cdot 17 \cdot 19 \cdot 29 \cdot 31 \cdot 53 \cdot 59 \cdot 61$

$648\,045\,936\,942\,300 = 2^2 \cdot 3^2 \cdot 5^2 \cdot 13 \cdot 17 \cdot 19 \cdot 29 \cdot 31 \cdot 53 \cdot 59 \cdot 61$

$1\,867\,897\,112\,363\,100 = 2^2 \cdot 3^2 \cdot 5^2 \cdot 7^2 \cdot 13 \cdot 19 \cdot 29 \cdot 31 \cdot 53 \cdot 59 \cdot 61$

$4\,981\,058\,966\,301\,600 = 2^5 \cdot 3 \cdot 5^2 \cdot 7^2 \cdot 13 \cdot 19 \cdot 29 \cdot 31 \cdot 53 \cdot 59 \cdot 61$

$12\,321\,566\,916\,640\,800 = 2^5 \cdot 3 \cdot 5^2 \cdot 7^2 \cdot 13 \cdot 29 \cdot 31 \cdot 47 \cdot 53 \cdot 59 \cdot 61$

$28\,339\,603\,908\,273\,840 = 2^4 \cdot 3 \cdot 5 \cdot 7^2 \cdot 13 \cdot 23 \cdot 29 \cdot 31 \cdot 47 \cdot 53 \cdot 59 \cdot 61$

$60\,727\,722\,660\,586\,800 = 2^4 \cdot 3^2 \cdot 5^2 \cdot 7 \cdot 13 \cdot 23 \cdot 29 \cdot 31 \cdot 47 \cdot 53 \cdot 59 \cdot 61$

$121\,455\,445\,321\,173\,600 = 2^5 \cdot 3^2 \cdot 5^2 \cdot 7 \cdot 13 \cdot 23 \cdot 29 \cdot 31 \cdot 47 \cdot 53 \cdot 59 \cdot 61$

$227\,068\,876\,035\,237\,600 = 2^5 \cdot 3^2 \cdot 5^2 \cdot 7 \cdot 13 \cdot 29 \cdot 31 \cdot 43 \cdot 47 \cdot 53 \cdot 59 \cdot 61$

$397\,370\,533\,061\,665\,800 = 2^3 \cdot 3^2 \cdot 5^2 \cdot 7^2 \cdot 13 \cdot 29 \cdot 31 \cdot 43 \cdot 47 \cdot 53 \cdot 59 \cdot 61$

$651\,687\,674\,221\,131\,912 = 2^3 \cdot 3^2 \cdot 7^2 \cdot 13 \cdot 29 \cdot 31 \cdot 41 \cdot 43 \cdot 47 \cdot 53 \cdot 59 \cdot 61$

$1\,002\,596\,421\,878\,664\,480 = 2^5 \cdot 3^2 \cdot 5 \cdot 7^2 \cdot 29 \cdot 31 \cdot 41 \cdot 43 \cdot 47 \cdot 53 \cdot 59 \cdot 61$

$1\,448\,194\,831\,602\,515\,360 = 2^5 \cdot 5 \cdot 7^2 \cdot 13 \cdot 29 \cdot 31 \cdot 41 \cdot 43 \cdot 47 \cdot 53 \cdot 59 \cdot 61$

$1\,965\,407\,271\,460\,556\,560 = 2^4 \cdot 5 \cdot 7 \cdot 13 \cdot 19 \cdot 29 \cdot 31 \cdot 41 \cdot 43 \cdot 47 \cdot 53 \cdot 59 \cdot 61$

$2\,507\,588\,587\,725\,537\,680 = 2^4 \cdot 5 \cdot 7 \cdot 13 \cdot 19 \cdot 31 \cdot 37 \cdot 41 \cdot 43 \cdot 47 \cdot 53 \cdot 59 \cdot 61$

$3\,009\,106\,305\,270\,645\,216 = 2^5 \cdot 3 \cdot 7 \cdot 13 \cdot 19 \cdot 31 \cdot 37 \cdot 41 \cdot 43 \cdot 47 \cdot 53 \cdot 59 \cdot 61$

$3\,397\,378\,086\,595\,889\,760 = 2^5 \cdot 3 \cdot 5 \cdot 7^2 \cdot 13 \cdot 19 \cdot 37 \cdot 41 \cdot 43 \cdot 47 \cdot 53 \cdot 59 \cdot 61$

$3\,609\,714\,217\,008\,132\,870 = 2 \cdot 3 \cdot 5 \cdot 7^2 \cdot 13 \cdot 17 \cdot 19 \cdot 37 \cdot 41 \cdot 43 \cdot 47 \cdot 53 \cdot 59 \cdot 61$

Row 66

$$66 = 2 \cdot 3 \cdot 11$$

$$2\,145 = 3 \cdot 5 \cdot 11 \cdot 13$$

$$45\,760 = 2^6 \cdot 5 \cdot 11 \cdot 13$$

$$720\,720 = 2^4 \cdot 3^2 \cdot 5 \cdot 7 \cdot 11 \cdot 13$$

$$8\,936\,928 = 2^5 \cdot 3^2 \cdot 7 \cdot 11 \cdot 13 \cdot 31$$

$$90\,858\,768 = 2^4 \cdot 3 \cdot 7 \cdot 11 \cdot 13 \cdot 31 \cdot 61$$

$$778\,789\,440 = 2^6 \cdot 3^2 \cdot 5 \cdot 11 \cdot 13 \cdot 31 \cdot 61$$

$$5\,743\,572\,120 = 2^3 \cdot 3^2 \cdot 5 \cdot 11 \cdot 13 \cdot 31 \cdot 59 \cdot 61$$

$$37\,014\,131\,440 = 2^4 \cdot 5 \cdot 11 \cdot 13 \cdot 29 \cdot 31 \cdot 59 \cdot 61$$

$$210\,980\,549\,208 = 2^3 \cdot 3 \cdot 11 \cdot 13 \cdot 19 \cdot 29 \cdot 31 \cdot 59 \cdot 61$$

$$1\,074\,082\,795\,968 = 2^6 \cdot 3 \cdot 7 \cdot 13 \cdot 19 \cdot 29 \cdot 31 \cdot 59 \cdot 61$$

$$4\,922\,879\,481\,520 = 2^4 \cdot 5 \cdot 7 \cdot 11 \cdot 13 \cdot 19 \cdot 29 \cdot 31 \cdot 59 \cdot 61$$

$$20\,448\,884\,000\,160 = 2^5 \cdot 3^3 \cdot 5 \cdot 7 \cdot 11 \cdot 19 \cdot 29 \cdot 31 \cdot 59 \cdot 61$$

$$77\,413\,632\,286\,320 = 2^4 \cdot 3^3 \cdot 5 \cdot 11 \cdot 19 \cdot 29 \cdot 31 \cdot 53 \cdot 59 \cdot 61$$

$$268\,367\,258\,592\,576 = 2^6 \cdot 3^2 \cdot 11 \cdot 13 \cdot 19 \cdot 29 \cdot 31 \cdot 53 \cdot 59 \cdot 61$$

$$855\,420\,636\,763\,836 = 2^2 \cdot 3^3 \cdot 11 \cdot 13 \cdot 17 \cdot 19 \cdot 29 \cdot 31 \cdot 53 \cdot 59 \cdot 61$$

$$2\,515\,943\,049\,305\,400 = 2^3 \cdot 3^3 \cdot 5^2 \cdot 11 \cdot 13 \cdot 19 \cdot 29 \cdot 31 \cdot 53 \cdot 59 \cdot 61$$

$$6\,848\,956\,078\,664\,700 = 2^2 \cdot 3 \cdot 5^2 \cdot 7^2 \cdot 11 \cdot 13 \cdot 19 \cdot 29 \cdot 31 \cdot 53 \cdot 59 \cdot 61$$

$$17\,302\,625\,882\,942\,400 = 2^6 \cdot 3^2 \cdot 5^2 \cdot 7^2 \cdot 11 \cdot 13 \cdot 29 \cdot 31 \cdot 53 \cdot 59 \cdot 61$$

$$40\,661\,170\,824\,914\,640 = 2^4 \cdot 3^2 \cdot 5 \cdot 7^2 \cdot 11 \cdot 13 \cdot 29 \cdot 31 \cdot 47 \cdot 53 \cdot 59 \cdot 61$$

$$89\,067\,326\,568\,860\,640 = 2^5 \cdot 3 \cdot 5 \cdot 7 \cdot 11 \cdot 13 \cdot 23 \cdot 29 \cdot 31 \cdot 47 \cdot 53 \cdot 59 \cdot 61$$

$$182\,183\,167\,981\,760\,400 = 2^4 \cdot 3^3 \cdot 5^2 \cdot 7 \cdot 13 \cdot 23 \cdot 29 \cdot 31 \cdot 47 \cdot 53 \cdot 59 \cdot 61$$

$$348\,524\,321\,356\,411\,200 = 2^6 \cdot 3^3 \cdot 5^2 \cdot 7 \cdot 11 \cdot 13 \cdot 29 \cdot 31 \cdot 47 \cdot 53 \cdot 59 \cdot 61$$

$$624\,439\,409\,096\,903\,400 = 2^3 \cdot 3^2 \cdot 5^2 \cdot 7 \cdot 11 \cdot 13 \cdot 29 \cdot 31 \cdot 43 \cdot 47 \cdot 53 \cdot 59 \cdot 61$$

$$1\,049\,058\,207\,282\,797\,712 = 2^4 \cdot 3^3 \cdot 7^2 \cdot 11 \cdot 13 \cdot 29 \cdot 31 \cdot 43 \cdot 47 \cdot 53 \cdot 59 \cdot 61$$

$$1\,654\,284\,096\,099\,796\,392 = 2^3 \cdot 3^3 \cdot 7^2 \cdot 11 \cdot 29 \cdot 31 \cdot 41 \cdot 43 \cdot 47 \cdot 53 \cdot 59 \cdot 61$$

$$2\,450\,791\,253\,481\,179\,840 = 2^6 \cdot 5 \cdot 7^2 \cdot 11 \cdot 29 \cdot 31 \cdot 41 \cdot 43 \cdot 47 \cdot 53 \cdot 59 \cdot 61$$

$$3\,413\,602\,103\,063\,071\,920 = 2^4 \cdot 3 \cdot 5 \cdot 7 \cdot 11 \cdot 13 \cdot 29 \cdot 31 \cdot 41 \cdot 43 \cdot 47 \cdot 53 \cdot 59 \cdot 61$$

$$4\,472\,995\,859\,186\,094\,240 = 2^5 \cdot 3 \cdot 5 \cdot 7 \cdot 11 \cdot 13 \cdot 19 \cdot 31 \cdot 41 \cdot 43 \cdot 47 \cdot 53 \cdot 59 \cdot 61$$

$$5\,516\,694\,892\,996\,182\,896 = 2^4 \cdot 7 \cdot 11 \cdot 13 \cdot 19 \cdot 31 \cdot 37 \cdot 41 \cdot 43 \cdot 47 \cdot 53 \cdot 59 \cdot 61$$

$$6\,406\,484\,391\,866\,534\,976 = 2^6 \cdot 3^2 \cdot 7 \cdot 11 \cdot 13 \cdot 19 \cdot 37 \cdot 41 \cdot 43 \cdot 47 \cdot 53 \cdot 59 \cdot 61$$

$$7\,007\,092\,303\,604\,022\,630 = 2 \cdot 3^2 \cdot 5 \cdot 7^2 \cdot 11 \cdot 13 \cdot 19 \cdot 37 \cdot 41 \cdot 43 \cdot 47 \cdot 53 \cdot 59 \cdot 61$$

$$7\,219\,428\,434\,016\,265\,740 = 2^2 \cdot 3 \cdot 5 \cdot 7^2 \cdot 13 \cdot 17 \cdot 19 \cdot 37 \cdot 41 \cdot 43 \cdot 47 \cdot 53 \cdot 59 \cdot 61$$

Row 67

$$67 \text{ is Prime}$$
$$2\,211 = 3 \cdot 11 \cdot 67$$
$$47\,905 = 5 \cdot 11 \cdot 13 \cdot 67$$
$$766\,480 = 2^4 \cdot 5 \cdot 11 \cdot 13 \cdot 67$$
$$9\,657\,648 = 2^4 \cdot 3^2 \cdot 7 \cdot 11 \cdot 13 \cdot 67$$
$$99\,795\,696 = 2^4 \cdot 3 \cdot 7 \cdot 11 \cdot 13 \cdot 31 \cdot 67$$
$$869\,648\,208 = 2^4 \cdot 3 \cdot 11 \cdot 13 \cdot 31 \cdot 61 \cdot 67$$
$$6\,522\,361\,560 = 2^3 \cdot 3^2 \cdot 5 \cdot 11 \cdot 13 \cdot 31 \cdot 61 \cdot 67$$
$$42\,757\,703\,560 = 2^3 \cdot 5 \cdot 11 \cdot 13 \cdot 31 \cdot 59 \cdot 61 \cdot 67$$
$$247\,994\,680\,648 = 2^3 \cdot 11 \cdot 13 \cdot 29 \cdot 31 \cdot 59 \cdot 61 \cdot 67$$
$$1\,285\,063\,345\,176 = 2^3 \cdot 3 \cdot 13 \cdot 19 \cdot 29 \cdot 31 \cdot 59 \cdot 61 \cdot 67$$
$$5\,996\,962\,277\,488 = 2^4 \cdot 7 \cdot 13 \cdot 19 \cdot 29 \cdot 31 \cdot 59 \cdot 61 \cdot 67$$
$$25\,371\,763\,481\,680 = 2^4 \cdot 5 \cdot 7 \cdot 11 \cdot 19 \cdot 29 \cdot 31 \cdot 59 \cdot 61 \cdot 67$$
$$97\,862\,516\,286\,480 = 2^4 \cdot 3^3 \cdot 5 \cdot 11 \cdot 19 \cdot 29 \cdot 31 \cdot 59 \cdot 61 \cdot 67$$
$$345\,780\,890\,878\,896 = 2^4 \cdot 3^2 \cdot 11 \cdot 19 \cdot 29 \cdot 31 \cdot 53 \cdot 59 \cdot 61 \cdot 67$$
$$1\,123\,787\,895\,356\,412 = 2^2 \cdot 3^2 \cdot 11 \cdot 13 \cdot 19 \cdot 29 \cdot 31 \cdot 53 \cdot 59 \cdot 61 \cdot 67$$
$$3\,371\,363\,686\,069\,236 = 2^2 \cdot 3^3 \cdot 11 \cdot 13 \cdot 19 \cdot 29 \cdot 31 \cdot 53 \cdot 59 \cdot 61 \cdot 67$$
$$9\,364\,899\,127\,970\,100 = 2^2 \cdot 3 \cdot 5^2 \cdot 11 \cdot 13 \cdot 19 \cdot 29 \cdot 31 \cdot 53 \cdot 59 \cdot 61 \cdot 67$$
$$24\,151\,581\,961\,607\,100 = 2^2 \cdot 3 \cdot 5^2 \cdot 7^2 \cdot 11 \cdot 13 \cdot 29 \cdot 31 \cdot 53 \cdot 59 \cdot 61 \cdot 67$$
$$57\,963\,796\,707\,857\,040 = 2^4 \cdot 3^2 \cdot 5 \cdot 7^2 \cdot 11 \cdot 13 \cdot 29 \cdot 31 \cdot 53 \cdot 59 \cdot 61 \cdot 67$$
$$129\,728\,497\,393\,775\,280 = 2^4 \cdot 3 \cdot 5 \cdot 7 \cdot 11 \cdot 13 \cdot 29 \cdot 31 \cdot 47 \cdot 53 \cdot 59 \cdot 61 \cdot 67$$
$$271\,250\,494\,550\,621\,040 = 2^4 \cdot 3 \cdot 5 \cdot 7 \cdot 13 \cdot 23 \cdot 29 \cdot 31 \cdot 47 \cdot 53 \cdot 59 \cdot 61 \cdot 67$$
$$530\,707\,489\,338\,171\,600 = 2^4 \cdot 3^3 \cdot 5^2 \cdot 7 \cdot 13 \cdot 29 \cdot 31 \cdot 47 \cdot 53 \cdot 59 \cdot 61 \cdot 67$$
$$972\,963\,730\,453\,314\,600 = 2^3 \cdot 3^2 \cdot 5^2 \cdot 7 \cdot 11 \cdot 13 \cdot 29 \cdot 31 \cdot 47 \cdot 53 \cdot 59 \cdot 61 \cdot 67$$
$$1\,673\,497\,616\,379\,701\,112 = 2^3 \cdot 3^2 \cdot 7 \cdot 11 \cdot 13 \cdot 29 \cdot 31 \cdot 43 \cdot 47 \cdot 53 \cdot 59 \cdot 61 \cdot 67$$
$$2\,703\,342\,303\,382\,594\,104 = 2^3 \cdot 3^3 \cdot 7^2 \cdot 11 \cdot 29 \cdot 31 \cdot 43 \cdot 47 \cdot 53 \cdot 59 \cdot 61 \cdot 67$$
$$4\,105\,075\,349\,580\,976\,232 = 2^3 \cdot 7^2 \cdot 11 \cdot 29 \cdot 31 \cdot 41 \cdot 43 \cdot 47 \cdot 53 \cdot 59 \cdot 61 \cdot 67$$
$$5\,864\,393\,356\,544\,251\,760 = 2^4 \cdot 5 \cdot 7 \cdot 11 \cdot 29 \cdot 31 \cdot 41 \cdot 43 \cdot 47 \cdot 53 \cdot 59 \cdot 61 \cdot 67$$
$$7\,886\,597\,962\,249\,166\,160 = 2^4 \cdot 3 \cdot 5 \cdot 7 \cdot 11 \cdot 13 \cdot 31 \cdot 41 \cdot 43 \cdot 47 \cdot 53 \cdot 59 \cdot 61 \cdot 67$$
$$9\,989\,690\,752\,182\,277\,136 = 2^4 \cdot 7 \cdot 11 \cdot 13 \cdot 19 \cdot 31 \cdot 41 \cdot 43 \cdot 47 \cdot 53 \cdot 59 \cdot 61 \cdot 67$$
$$11\,923\,179\,284\,862\,717\,872 = 2^4 \cdot 7 \cdot 11 \cdot 13 \cdot 19 \cdot 37 \cdot 41 \cdot 43 \cdot 47 \cdot 53 \cdot 59 \cdot 61 \cdot 67$$
$$13\,413\,576\,695\,470\,557\,606 = 2 \cdot 3^2 \cdot 7 \cdot 11 \cdot 13 \cdot 19 \cdot 37 \cdot 41 \cdot 43 \cdot 47 \cdot 53 \cdot 59 \cdot 61 \cdot 67$$
$$14\,226\,520\,737\,620\,288\,370 = 2 \cdot 3 \cdot 5 \cdot 7^2 \cdot 13 \cdot 19 \cdot 37 \cdot 41 \cdot 43 \cdot 47 \cdot 53 \cdot 59 \cdot 61 \cdot 67$$

<u>Row 68</u>

$$68 = 2^2 \cdot 17$$

$$2\,278 = 2 \cdot 17 \cdot 67$$

$$50\,116 = 2^2 \cdot 11 \cdot 17 \cdot 67$$

$$814\,385 = 5 \cdot 11 \cdot 13 \cdot 17 \cdot 67$$

$$10\,424\,128 = 2^6 \cdot 11 \cdot 13 \cdot 17 \cdot 67$$

$$109\,453\,344 = 2^5 \cdot 3 \cdot 7 \cdot 11 \cdot 13 \cdot 17 \cdot 67$$

$$969\,443\,904 = 2^6 \cdot 3 \cdot 11 \cdot 13 \cdot 17 \cdot 31 \cdot 67$$

$$7\,392\,009\,768 = 2^3 \cdot 3 \cdot 11 \cdot 13 \cdot 17 \cdot 31 \cdot 61 \cdot 67$$

$$49\,280\,065\,120 = 2^5 \cdot 5 \cdot 11 \cdot 13 \cdot 17 \cdot 31 \cdot 61 \cdot 67$$

$$290\,752\,384\,208 = 2^4 \cdot 11 \cdot 13 \cdot 17 \cdot 31 \cdot 59 \cdot 61 \cdot 67$$

$$1\,533\,058\,025\,824 = 2^5 \cdot 13 \cdot 17 \cdot 29 \cdot 31 \cdot 59 \cdot 61 \cdot 67$$

$$7\,282\,025\,622\,664 = 2^3 \cdot 13 \cdot 17 \cdot 19 \cdot 29 \cdot 31 \cdot 59 \cdot 61 \cdot 67$$

$$31\,368\,725\,759\,168 = 2^6 \cdot 7 \cdot 17 \cdot 19 \cdot 29 \cdot 31 \cdot 59 \cdot 61 \cdot 67$$

$$123\,234\,279\,768\,160 = 2^5 \cdot 5 \cdot 11 \cdot 17 \cdot 19 \cdot 29 \cdot 31 \cdot 59 \cdot 61 \cdot 67$$

$$443\,643\,407\,165\,376 = 2^6 \cdot 3^2 \cdot 11 \cdot 17 \cdot 19 \cdot 29 \cdot 31 \cdot 59 \cdot 61 \cdot 67$$

$$1\,469\,568\,786\,235\,308 = 2^2 \cdot 3^2 \cdot 11 \cdot 17 \cdot 19 \cdot 29 \cdot 31 \cdot 53 \cdot 59 \cdot 61 \cdot 67$$

$$4\,495\,151\,581\,425\,648 = 2^4 \cdot 3^2 \cdot 11 \cdot 13 \cdot 19 \cdot 29 \cdot 31 \cdot 53 \cdot 59 \cdot 61 \cdot 67$$

$$12\,736\,262\,814\,039\,336 = 2^3 \cdot 3 \cdot 11 \cdot 13 \cdot 17 \cdot 19 \cdot 31 \cdot 53 \cdot 59 \cdot 61 \cdot 67$$

$$33\,516\,481\,089\,577\,200 = 2^4 \cdot 3 \cdot 5^2 \cdot 11 \cdot 13 \cdot 17 \cdot 29 \cdot 31 \cdot 53 \cdot 59 \cdot 61 \cdot 67$$

$$82\,115\,378\,669\,464\,140 = 2^2 \cdot 3 \cdot 5 \cdot 7^2 \cdot 11 \cdot 13 \cdot 17 \cdot 29 \cdot 31 \cdot 53 \cdot 59 \cdot 61 \cdot 67$$

$$187\,692\,294\,101\,632\,320 = 2^6 \cdot 3 \cdot 5 \cdot 7 \cdot 11 \cdot 13 \cdot 17 \cdot 29 \cdot 31 \cdot 53 \cdot 59 \cdot 61 \cdot 67$$

$$400\,978\,991\,944\,396\,320 = 2^5 \cdot 3 \cdot 5 \cdot 7 \cdot 13 \cdot 17 \cdot 29 \cdot 31 \cdot 47 \cdot 53 \cdot 59 \cdot 61 \cdot 67$$

$$801\,957\,983\,888\,792\,640 = 2^6 \cdot 3 \cdot 5 \cdot 7 \cdot 13 \cdot 17 \cdot 29 \cdot 31 \cdot 47 \cdot 53 \cdot 59 \cdot 61 \cdot 67$$

$$1\,503\,671\,219\,791\,486\,200 = 2^3 \cdot 3^2 \cdot 5^2 \cdot 7 \cdot 13 \cdot 17 \cdot 29 \cdot 31 \cdot 47 \cdot 53 \cdot 59 \cdot 61 \cdot 67$$

$$2\,646\,461\,346\,833\,015\,712 = 2^5 \cdot 3^2 \cdot 7 \cdot 11 \cdot 13 \cdot 17 \cdot 29 \cdot 31 \cdot 47 \cdot 53 \cdot 59 \cdot 61 \cdot 67$$

$$4\,376\,839\,919\,762\,295\,216 = 2^4 \cdot 3^2 \cdot 7 \cdot 11 \cdot 17 \cdot 29 \cdot 31 \cdot 43 \cdot 47 \cdot 53 \cdot 59 \cdot 61 \cdot 67$$

$$6\,808\,417\,652\,963\,570\,336 = 2^5 \cdot 7^2 \cdot 11 \cdot 17 \cdot 29 \cdot 31 \cdot 43 \cdot 47 \cdot 53 \cdot 59 \cdot 61 \cdot 67$$

$$9\,969\,468\,706\,125\,227\,992 = 2^3 \cdot 7 \cdot 11 \cdot 17 \cdot 29 \cdot 31 \cdot 41 \cdot 43 \cdot 47 \cdot 53 \cdot 59 \cdot 61 \cdot 67$$

$$13\,750\,991\,318\,793\,417\,920 = 2^6 \cdot 5 \cdot 7 \cdot 11 \cdot 17 \cdot 31 \cdot 41 \cdot 43 \cdot 47 \cdot 53 \cdot 59 \cdot 61 \cdot 67$$

$$17\,876\,288\,714\,431\,443\,296 = 2^5 \cdot 7 \cdot 11 \cdot 13 \cdot 17 \cdot 31 \cdot 41 \cdot 43 \cdot 47 \cdot 53 \cdot 59 \cdot 61 \cdot 67$$

$$21\,912\,870\,037\,044\,995\,008 = 2^6 \cdot 7 \cdot 11 \cdot 13 \cdot 17 \cdot 19 \cdot 41 \cdot 43 \cdot 47 \cdot 53 \cdot 59 \cdot 61 \cdot 67$$

$$25\,336\,755\,980\,333\,275\,478 = 2 \cdot 7 \cdot 11 \cdot 13 \cdot 17 \cdot 19 \cdot 37 \cdot 41 \cdot 43 \cdot 47 \cdot 53 \cdot 59 \cdot 61 \cdot 67$$

$$27\,640\,097\,433\,090\,845\,976 = 2^3 \cdot 3 \cdot 7 \cdot 13 \cdot 17 \cdot 19 \cdot 37 \cdot 41 \cdot 43 \cdot 47 \cdot 53 \cdot 59 \cdot 61 \cdot 67$$

$$28\,453\,041\,475\,240\,576\,740 = 2^2 \cdot 3 \cdot 5 \cdot 7^2 \cdot 13 \cdot 19 \cdot 37 \cdot 41 \cdot 43 \cdot 47 \cdot 53 \cdot 59 \cdot 61 \cdot 67$$

Row 69

$$69 = 3 \cdot 23$$
$$2\,346 = 2 \cdot 3 \cdot 17 \cdot 23$$
$$52\,394 = 2 \cdot 17 \cdot 23 \cdot 67$$
$$864\,501 = 3 \cdot 11 \cdot 17 \cdot 23 \cdot 67$$
$$11\,238\,513 = 3 \cdot 11 \cdot 13 \cdot 17 \cdot 23 \cdot 67$$
$$119\,877\,472 = 2^5 \cdot 11 \cdot 13 \cdot 17 \cdot 23 \cdot 67$$
$$1\,078\,897\,248 = 2^5 \cdot 3^2 \cdot 11 \cdot 13 \cdot 17 \cdot 23 \cdot 67$$
$$8\,361\,453\,672 = 2^3 \cdot 3^2 \cdot 11 \cdot 13 \cdot 17 \cdot 23 \cdot 31 \cdot 67$$
$$56\,672\,074\,888 = 2^3 \cdot 11 \cdot 13 \cdot 17 \cdot 23 \cdot 31 \cdot 61 \cdot 67$$
$$340\,032\,449\,328 = 2^4 \cdot 3 \cdot 11 \cdot 13 \cdot 17 \cdot 23 \cdot 31 \cdot 61 \cdot 67$$
$$1\,823\,810\,410\,032 = 2^4 \cdot 3 \cdot 13 \cdot 17 \cdot 23 \cdot 31 \cdot 59 \cdot 61 \cdot 67$$
$$8\,815\,083\,648\,488 = 2^3 \cdot 13 \cdot 17 \cdot 23 \cdot 29 \cdot 31 \cdot 59 \cdot 61 \cdot 67$$
$$38\,650\,751\,381\,832 = 2^3 \cdot 3 \cdot 17 \cdot 19 \cdot 23 \cdot 29 \cdot 31 \cdot 59 \cdot 61 \cdot 67$$
$$154\,603\,005\,527\,328 = 2^5 \cdot 3 \cdot 17 \cdot 19 \cdot 23 \cdot 29 \cdot 31 \cdot 59 \cdot 61 \cdot 67$$
$$566\,877\,686\,933\,536 = 2^5 \cdot 11 \cdot 17 \cdot 19 \cdot 23 \cdot 29 \cdot 31 \cdot 59 \cdot 61 \cdot 67$$
$$1\,913\,212\,193\,400\,684 = 2^2 \cdot 3^3 \cdot 11 \cdot 17 \cdot 19 \cdot 23 \cdot 29 \cdot 31 \cdot 59 \cdot 61 \cdot 67$$
$$5\,964\,720\,367\,660\,956 = 2^2 \cdot 3^3 \cdot 11 \cdot 19 \cdot 23 \cdot 29 \cdot 31 \cdot 53 \cdot 59 \cdot 61 \cdot 67$$
$$17\,231\,414\,395\,464\,984 = 2^3 \cdot 3 \cdot 11 \cdot 13 \cdot 19 \cdot 23 \cdot 29 \cdot 31 \cdot 53 \cdot 59 \cdot 61 \cdot 67$$
$$46\,252\,743\,903\,616\,536 = 2^3 \cdot 3^2 \cdot 11 \cdot 13 \cdot 17 \cdot 23 \cdot 29 \cdot 31 \cdot 53 \cdot 59 \cdot 61 \cdot 67$$
$$115\,631\,859\,759\,041\,340 = 2^2 \cdot 3^2 \cdot 5 \cdot 11 \cdot 13 \cdot 17 \cdot 23 \cdot 29 \cdot 31 \cdot 53 \cdot 59 \cdot 61 \cdot 67$$
$$269\,807\,672\,771\,096\,460 = 2^2 \cdot 3 \cdot 5 \cdot 7 \cdot 11 \cdot 13 \cdot 17 \cdot 23 \cdot 29 \cdot 31 \cdot 53 \cdot 59 \cdot 61 \cdot 67$$
$$588\,671\,286\,046\,028\,640 = 2^5 \cdot 3^2 \cdot 5 \cdot 7 \cdot 13 \cdot 17 \cdot 23 \cdot 29 \cdot 31 \cdot 53 \cdot 59 \cdot 61 \cdot 67$$
$$1\,202\,936\,975\,833\,188\,960 = 2^5 \cdot 3^2 \cdot 5 \cdot 7 \cdot 13 \cdot 17 \cdot 29 \cdot 31 \cdot 47 \cdot 53 \cdot 59 \cdot 61 \cdot 67$$
$$2\,305\,629\,203\,680\,278\,840 = 2^3 \cdot 3 \cdot 5 \cdot 7 \cdot 13 \cdot 17 \cdot 23 \cdot 29 \cdot 31 \cdot 47 \cdot 53 \cdot 59 \cdot 61 \cdot 67$$
$$4\,150\,132\,566\,624\,501\,912 = 2^3 \cdot 3^3 \cdot 7 \cdot 13 \cdot 17 \cdot 23 \cdot 29 \cdot 31 \cdot 47 \cdot 53 \cdot 59 \cdot 61 \cdot 67$$
$$7\,023\,301\,266\,595\,310\,928 = 2^4 \cdot 3^3 \cdot 7 \cdot 11 \cdot 17 \cdot 23 \cdot 29 \cdot 31 \cdot 47 \cdot 53 \cdot 59 \cdot 61 \cdot 67$$
$$11\,185\,257\,572\,725\,865\,552 = 2^4 \cdot 7 \cdot 11 \cdot 17 \cdot 23 \cdot 29 \cdot 31 \cdot 43 \cdot 47 \cdot 53 \cdot 59 \cdot 61 \cdot 67$$
$$16\,777\,886\,359\,088\,798\,328 = 2^3 \cdot 3 \cdot 7 \cdot 11 \cdot 17 \cdot 23 \cdot 29 \cdot 31 \cdot 43 \cdot 47 \cdot 53 \cdot 59 \cdot 61 \cdot 67$$
$$23\,720\,460\,024\,918\,645\,912 = 2^3 \cdot 3 \cdot 7 \cdot 11 \cdot 17 \cdot 23 \cdot 31 \cdot 41 \cdot 43 \cdot 47 \cdot 53 \cdot 59 \cdot 61 \cdot 67$$
$$31\,627\,280\,033\,224\,861\,216 = 2^5 \cdot 7 \cdot 11 \cdot 17 \cdot 23 \cdot 31 \cdot 41 \cdot 43 \cdot 47 \cdot 53 \cdot 59 \cdot 61 \cdot 67$$
$$39\,789\,158\,751\,476\,438\,304 = 2^5 \cdot 3 \cdot 7 \cdot 11 \cdot 13 \cdot 17 \cdot 23 \cdot 41 \cdot 43 \cdot 47 \cdot 53 \cdot 59 \cdot 61 \cdot 67$$
$$47\,249\,626\,017\,378\,270\,486 = 2 \cdot 3 \cdot 7 \cdot 11 \cdot 13 \cdot 17 \cdot 19 \cdot 23 \cdot 41 \cdot 43 \cdot 47 \cdot 53 \cdot 59 \cdot 61 \cdot 67$$
$$52\,976\,853\,413\,424\,121\,454 = 2 \cdot 7 \cdot 13 \cdot 17 \cdot 19 \cdot 23 \cdot 37 \cdot 41 \cdot 43 \cdot 47 \cdot 53 \cdot 59 \cdot 61 \cdot 67$$
$$56\,093\,138\,908\,331\,422\,716 = 2^2 \cdot 3^2 \cdot 7 \cdot 13 \cdot 19 \cdot 23 \cdot 37 \cdot 41 \cdot 43 \cdot 47 \cdot 53 \cdot 59 \cdot 61 \cdot 67$$

<u>Row 70</u>

$70 = 2 \cdot 5 \cdot 7$

$2\,415 = 3 \cdot 5 \cdot 7 \cdot 23$

$54\,740 = 2^2 \cdot 5 \cdot 7 \cdot 17 \cdot 23$

$916\,895 = 5 \cdot 7 \cdot 17 \cdot 23 \cdot 67$

$12\,103\,014 = 2 \cdot 3 \cdot 7 \cdot 11 \cdot 17 \cdot 23 \cdot 67$

$131\,115\,985 = 5 \cdot 7 \cdot 11 \cdot 13 \cdot 17 \cdot 23 \cdot 67$

$1\,198\,774\,720 = 2^6 \cdot 5 \cdot 11 \cdot 13 \cdot 17 \cdot 23 \cdot 67$

$9\,440\,350\,920 = 2^3 \cdot 3^2 \cdot 5 \cdot 7 \cdot 11 \cdot 13 \cdot 17 \cdot 23 \cdot 67$

$65\,033\,528\,560 = 2^4 \cdot 5 \cdot 7 \cdot 11 \cdot 13 \cdot 17 \cdot 23 \cdot 31 \cdot 67$

$396\,704\,524\,216 = 2^3 \cdot 7 \cdot 11 \cdot 13 \cdot 17 \cdot 23 \cdot 31 \cdot 61 \cdot 67$

$2\,163\,842\,859\,360 = 2^5 \cdot 3 \cdot 5 \cdot 7 \cdot 13 \cdot 17 \cdot 23 \cdot 31 \cdot 61 \cdot 67$

$10\,638\,894\,058\,520 = 2^3 \cdot 5 \cdot 7 \cdot 13 \cdot 17 \cdot 23 \cdot 31 \cdot 59 \cdot 61 \cdot 67$

$47\,465\,835\,030\,320 = 2^4 \cdot 5 \cdot 7 \cdot 17 \cdot 23 \cdot 29 \cdot 31 \cdot 59 \cdot 61 \cdot 67$

$193\,253\,756\,909\,160 = 2^3 \cdot 3 \cdot 5 \cdot 17 \cdot 19 \cdot 23 \cdot 29 \cdot 31 \cdot 59 \cdot 61 \cdot 67$

$721\,480\,692\,460\,864 = 2^6 \cdot 7 \cdot 17 \cdot 19 \cdot 23 \cdot 29 \cdot 31 \cdot 59 \cdot 61 \cdot 67$

$2\,480\,089\,880\,334\,220 = 2^2 \cdot 5 \cdot 7 \cdot 11 \cdot 17 \cdot 19 \cdot 23 \cdot 29 \cdot 31 \cdot 59 \cdot 61 \cdot 67$

$7\,877\,932\,561\,061\,640 = 2^3 \cdot 3^3 \cdot 5 \cdot 7 \cdot 11 \cdot 19 \cdot 23 \cdot 29 \cdot 31 \cdot 59 \cdot 61 \cdot 67$

$23\,196\,134\,763\,125\,940 = 2^2 \cdot 3 \cdot 5 \cdot 7 \cdot 11 \cdot 19 \cdot 23 \cdot 29 \cdot 31 \cdot 53 \cdot 59 \cdot 61 \cdot 67$

$63\,484\,158\,299\,081\,520 = 2^4 \cdot 3 \cdot 5 \cdot 7 \cdot 11 \cdot 13 \cdot 23 \cdot 29 \cdot 31 \cdot 53 \cdot 59 \cdot 61 \cdot 67$

$161\,884\,603\,662\,657\,876 = 2^2 \cdot 3^2 \cdot 7 \cdot 11 \cdot 13 \cdot 17 \cdot 23 \cdot 29 \cdot 31 \cdot 53 \cdot 59 \cdot 61 \cdot 67$

$385\,439\,532\,530\,137\,800 = 2^3 \cdot 3 \cdot 5^2 \cdot 11 \cdot 13 \cdot 17 \cdot 23 \cdot 29 \cdot 31 \cdot 53 \cdot 59 \cdot 61 \cdot 67$

$858\,478\,958\,817\,125\,100 = 2^2 \cdot 3 \cdot 5^2 \cdot 7^2 \cdot 13 \cdot 17 \cdot 23 \cdot 29 \cdot 31 \cdot 53 \cdot 59 \cdot 61 \cdot 67$

$1\,791\,608\,261\,879\,217\,600 = 2^6 \cdot 3^2 \cdot 5^2 \cdot 7^2 \cdot 13 \cdot 17 \cdot 29 \cdot 31 \cdot 53 \cdot 59 \cdot 61 \cdot 67$

$3\,508\,566\,179\,513\,467\,800 = 2^3 \cdot 3 \cdot 5^2 \cdot 7^2 \cdot 13 \cdot 17 \cdot 29 \cdot 31 \cdot 47 \cdot 53 \cdot 59 \cdot 61 \cdot 67$

$6\,455\,761\,770\,304\,780\,752 = 2^4 \cdot 3 \cdot 7^2 \cdot 13 \cdot 17 \cdot 23 \cdot 29 \cdot 31 \cdot 47 \cdot 53 \cdot 59 \cdot 61 \cdot 67$

$11\,173\,433\,833\,219\,812\,840 = 2^3 \cdot 3^3 \cdot 5 \cdot 7^2 \cdot 17 \cdot 23 \cdot 29 \cdot 31 \cdot 47 \cdot 53 \cdot 59 \cdot 61 \cdot 67$

$18\,208\,558\,839\,321\,176\,480 = 2^5 \cdot 5 \cdot 7^2 \cdot 11 \cdot 17 \cdot 23 \cdot 29 \cdot 31 \cdot 47 \cdot 53 \cdot 59 \cdot 61 \cdot 67$

$27\,963\,143\,931\,814\,663\,880 = 2^3 \cdot 5 \cdot 7 \cdot 11 \cdot 17 \cdot 23 \cdot 29 \cdot 31 \cdot 43 \cdot 47 \cdot 53 \cdot 59 \cdot 61 \cdot 67$

$40\,498\,346\,384\,007\,444\,240 = 2^4 \cdot 3 \cdot 5 \cdot 7^2 \cdot 11 \cdot 17 \cdot 23 \cdot 31 \cdot 43 \cdot 47 \cdot 53 \cdot 59 \cdot 61 \cdot 67$

$55\,347\,740\,058\,143\,507\,128 = 2^3 \cdot 7^2 \cdot 11 \cdot 17 \cdot 23 \cdot 31 \cdot 41 \cdot 43 \cdot 47 \cdot 53 \cdot 59 \cdot 61 \cdot 67$

$71\,416\,438\,784\,701\,299\,520 = 2^6 \cdot 5 \cdot 7^2 \cdot 11 \cdot 17 \cdot 23 \cdot 41 \cdot 43 \cdot 47 \cdot 53 \cdot 59 \cdot 61 \cdot 67$

$87\,038\,784\,768\,854\,708\,790 = 2 \cdot 3 \cdot 5 \cdot 7^2 \cdot 11 \cdot 13 \cdot 17 \cdot 23 \cdot 41 \cdot 43 \cdot 47 \cdot 53 \cdot 59 \cdot 61 \cdot 67$

$100\,226\,479\,430\,802\,391\,940 = 2^2 \cdot 5 \cdot 7^2 \cdot 13 \cdot 17 \cdot 19 \cdot 23 \cdot 41 \cdot 43 \cdot 47 \cdot 53 \cdot 59 \cdot 61 \cdot 67$

$109\,069\,992\,321\,755\,544\,170 = 2 \cdot 5 \cdot 7^2 \cdot 13 \cdot 19 \cdot 23 \cdot 37 \cdot 41 \cdot 43 \cdot 47 \cdot 53 \cdot 59 \cdot 61 \cdot 67$

$112\,186\,277\,816\,662\,845\,432 = 2^3 \cdot 3^2 \cdot 7 \cdot 13 \cdot 19 \cdot 23 \cdot 37 \cdot 41 \cdot 43 \cdot 47 \cdot 53 \cdot 59 \cdot 61 \cdot 67$

Row 71

$$71 \text{ is Prime}$$

$$2\,485 = 5 \cdot 7 \cdot 71$$

$$57\,155 = 5 \cdot 7 \cdot 23 \cdot 71$$

$$971\,635 = 5 \cdot 7 \cdot 17 \cdot 23 \cdot 71$$

$$13\,019\,909 = 7 \cdot 17 \cdot 23 \cdot 67 \cdot 71$$

$$143\,218\,999 = 7 \cdot 11 \cdot 17 \cdot 23 \cdot 67 \cdot 71$$

$$1\,329\,890\,705 = 5 \cdot 11 \cdot 13 \cdot 17 \cdot 23 \cdot 67 \cdot 71$$

$$10\,639\,125\,640 = 2^3 \cdot 5 \cdot 11 \cdot 13 \cdot 17 \cdot 23 \cdot 67 \cdot 71$$

$$74\,473\,879\,480 = 2^3 \cdot 5 \cdot 7 \cdot 11 \cdot 13 \cdot 17 \cdot 23 \cdot 67 \cdot 71$$

$$461\,738\,052\,776 = 2^3 \cdot 7 \cdot 11 \cdot 13 \cdot 17 \cdot 23 \cdot 31 \cdot 67 \cdot 71$$

$$2\,560\,547\,383\,576 = 2^3 \cdot 7 \cdot 13 \cdot 17 \cdot 23 \cdot 31 \cdot 61 \cdot 67 \cdot 71$$

$$12\,802\,736\,917\,880 = 2^3 \cdot 5 \cdot 7 \cdot 13 \cdot 17 \cdot 23 \cdot 31 \cdot 61 \cdot 67 \cdot 71$$

$$58\,104\,729\,088\,840 = 2^3 \cdot 5 \cdot 7 \cdot 17 \cdot 23 \cdot 31 \cdot 59 \cdot 61 \cdot 67 \cdot 71$$

$$240\,719\,591\,939\,480 = 2^3 \cdot 5 \cdot 17 \cdot 23 \cdot 29 \cdot 31 \cdot 59 \cdot 61 \cdot 67 \cdot 71$$

$$914\,734\,449\,370\,024 = 2^3 \cdot 17 \cdot 19 \cdot 23 \cdot 29 \cdot 31 \cdot 59 \cdot 61 \cdot 67 \cdot 71$$

$$3\,201\,570\,572\,795\,084 = 2^2 \cdot 7 \cdot 17 \cdot 19 \cdot 23 \cdot 29 \cdot 31 \cdot 59 \cdot 61 \cdot 67 \cdot 71$$

$$10\,358\,022\,441\,395\,860 = 2^2 \cdot 5 \cdot 7 \cdot 11 \cdot 19 \cdot 23 \cdot 29 \cdot 31 \cdot 59 \cdot 61 \cdot 67 \cdot 71$$

$$31\,074\,067\,324\,187\,580 = 2^2 \cdot 3 \cdot 5 \cdot 7 \cdot 11 \cdot 19 \cdot 23 \cdot 29 \cdot 31 \cdot 59 \cdot 61 \cdot 67 \cdot 71$$

$$86\,680\,293\,062\,207\,460 = 2^2 \cdot 3 \cdot 5 \cdot 7 \cdot 11 \cdot 23 \cdot 29 \cdot 31 \cdot 53 \cdot 59 \cdot 61 \cdot 67 \cdot 71$$

$$225\,368\,761\,961\,739\,396 = 2^2 \cdot 3 \cdot 7 \cdot 11 \cdot 13 \cdot 23 \cdot 29 \cdot 31 \cdot 53 \cdot 59 \cdot 61 \cdot 67 \cdot 71$$

$$547\,324\,136\,192\,795\,676 = 2^2 \cdot 3 \cdot 11 \cdot 13 \cdot 17 \cdot 23 \cdot 29 \cdot 31 \cdot 53 \cdot 59 \cdot 61 \cdot 67 \cdot 71$$

$$1\,243\,918\,491\,347\,262\,900 = 2^2 \cdot 3 \cdot 5^2 \cdot 13 \cdot 17 \cdot 23 \cdot 29 \cdot 31 \cdot 53 \cdot 59 \cdot 61 \cdot 67 \cdot 71$$

$$2\,650\,087\,220\,696\,342\,700 = 2^2 \cdot 3 \cdot 5^2 \cdot 7^2 \cdot 13 \cdot 17 \cdot 29 \cdot 31 \cdot 53 \cdot 59 \cdot 61 \cdot 67 \cdot 71$$

$$5\,300\,174\,441\,392\,685\,400 = 2^3 \cdot 3 \cdot 5^2 \cdot 7^2 \cdot 13 \cdot 17 \cdot 29 \cdot 31 \cdot 53 \cdot 59 \cdot 61 \cdot 67 \cdot 71$$

$$9\,964\,327\,949\,818\,248\,552 = 2^3 \cdot 3 \cdot 7^2 \cdot 13 \cdot 17 \cdot 29 \cdot 31 \cdot 47 \cdot 53 \cdot 59 \cdot 61 \cdot 67 \cdot 71$$

$$17\,629\,195\,603\,524\,593\,592 = 2^3 \cdot 3 \cdot 7^2 \cdot 17 \cdot 23 \cdot 29 \cdot 31 \cdot 47 \cdot 53 \cdot 59 \cdot 61 \cdot 67 \cdot 71$$

$$29\,381\,992\,672\,540\,989\,320 = 2^3 \cdot 5 \cdot 7^2 \cdot 17 \cdot 23 \cdot 29 \cdot 31 \cdot 47 \cdot 53 \cdot 59 \cdot 61 \cdot 67 \cdot 71$$

$$46\,171\,702\,771\,135\,840\,360 = 2^3 \cdot 5 \cdot 7 \cdot 11 \cdot 17 \cdot 23 \cdot 29 \cdot 31 \cdot 47 \cdot 53 \cdot 59 \cdot 61 \cdot 67 \cdot 71$$

$$68\,461\,490\,315\,822\,108\,120 = 2^3 \cdot 5 \cdot 7 \cdot 11 \cdot 17 \cdot 23 \cdot 31 \cdot 43 \cdot 47 \cdot 53 \cdot 59 \cdot 61 \cdot 67 \cdot 71$$

$$95\,846\,086\,442\,150\,951\,368 = 2^3 \cdot 7^2 \cdot 11 \cdot 17 \cdot 23 \cdot 31 \cdot 43 \cdot 47 \cdot 53 \cdot 59 \cdot 61 \cdot 67 \cdot 71$$

$$126\,764\,178\,842\,844\,806\,648 = 2^3 \cdot 7^2 \cdot 11 \cdot 17 \cdot 23 \cdot 41 \cdot 43 \cdot 47 \cdot 53 \cdot 59 \cdot 61 \cdot 67 \cdot 71$$

$$158\,455\,223\,553\,556\,008\,310 = 2 \cdot 5 \cdot 7^2 \cdot 11 \cdot 17 \cdot 23 \cdot 41 \cdot 43 \cdot 47 \cdot 53 \cdot 59 \cdot 61 \cdot 67 \cdot 71$$

$$187\,265\,264\,199\,657\,100\,730 = 2 \cdot 5 \cdot 7^2 \cdot 13 \cdot 17 \cdot 23 \cdot 41 \cdot 43 \cdot 47 \cdot 53 \cdot 59 \cdot 61 \cdot 67 \cdot 71$$

$$209\,296\,471\,752\,557\,936\,110 = 2 \cdot 5 \cdot 7^2 \cdot 13 \cdot 19 \cdot 23 \cdot 41 \cdot 43 \cdot 47 \cdot 53 \cdot 59 \cdot 61 \cdot 67 \cdot 71$$

$$221\,256\,270\,138\,418\,389\,602 = 2 \cdot 7 \cdot 13 \cdot 19 \cdot 23 \cdot 37 \cdot 41 \cdot 43 \cdot 47 \cdot 53 \cdot 59 \cdot 61 \cdot 67 \cdot 71$$

Row 72

$$72 = 2^3 \cdot 3^2$$
$$2\,556 = 2^2 \cdot 3^2 \cdot 71$$
$$59\,640 = 2^3 \cdot 3 \cdot 5 \cdot 7 \cdot 71$$
$$1\,028\,790 = 2 \cdot 3^2 \cdot 5 \cdot 7 \cdot 23 \cdot 71$$
$$13\,991\,544 = 2^3 \cdot 3^2 \cdot 7 \cdot 17 \cdot 23 \cdot 71$$
$$156\,238\,908 = 2^2 \cdot 3 \cdot 7 \cdot 17 \cdot 23 \cdot 67 \cdot 71$$
$$1\,473\,109\,704 = 2^3 \cdot 3^2 \cdot 11 \cdot 17 \cdot 23 \cdot 67 \cdot 71$$
$$11\,969\,016\,345 = 3^2 \cdot 5 \cdot 11 \cdot 13 \cdot 17 \cdot 23 \cdot 67 \cdot 71$$
$$85\,113\,005\,120 = 2^6 \cdot 5 \cdot 11 \cdot 13 \cdot 17 \cdot 23 \cdot 67 \cdot 71$$
$$536\,211\,932\,256 = 2^5 \cdot 3^2 \cdot 7 \cdot 11 \cdot 13 \cdot 17 \cdot 23 \cdot 67 \cdot 71$$
$$3\,022\,285\,436\,352 = 2^6 \cdot 3^2 \cdot 7 \cdot 13 \cdot 17 \cdot 23 \cdot 31 \cdot 67 \cdot 71$$
$$15\,363\,284\,301\,456 = 2^4 \cdot 3 \cdot 7 \cdot 13 \cdot 17 \cdot 23 \cdot 31 \cdot 61 \cdot 67 \cdot 71$$
$$70\,907\,466\,006\,720 = 2^6 \cdot 3^2 \cdot 5 \cdot 7 \cdot 17 \cdot 23 \cdot 31 \cdot 61 \cdot 67 \cdot 71$$
$$298\,824\,321\,028\,320 = 2^5 \cdot 3^2 \cdot 5 \cdot 17 \cdot 23 \cdot 31 \cdot 59 \cdot 61 \cdot 67 \cdot 71$$
$$1\,155\,454\,041\,309\,504 = 2^6 \cdot 3 \cdot 17 \cdot 23 \cdot 29 \cdot 31 \cdot 59 \cdot 61 \cdot 67 \cdot 71$$
$$4\,116\,305\,022\,165\,108 = 2^2 \cdot 3^2 \cdot 17 \cdot 19 \cdot 23 \cdot 29 \cdot 31 \cdot 59 \cdot 61 \cdot 67 \cdot 71$$
$$13\,559\,593\,014\,190\,944 = 2^5 \cdot 3^2 \cdot 7 \cdot 19 \cdot 23 \cdot 29 \cdot 31 \cdot 59 \cdot 61 \cdot 67 \cdot 71$$
$$41\,432\,089\,765\,583\,440 = 2^4 \cdot 5 \cdot 7 \cdot 11 \cdot 19 \cdot 23 \cdot 29 \cdot 31 \cdot 59 \cdot 61 \cdot 67 \cdot 71$$
$$117\,754\,360\,386\,395\,040 = 2^5 \cdot 3^3 \cdot 5 \cdot 7 \cdot 11 \cdot 23 \cdot 29 \cdot 31 \cdot 59 \cdot 61 \cdot 67 \cdot 71$$
$$312\,049\,055\,023\,946\,856 = 2^3 \cdot 3^3 \cdot 7 \cdot 11 \cdot 23 \cdot 29 \cdot 31 \cdot 53 \cdot 59 \cdot 61 \cdot 67 \cdot 71$$
$$772\,692\,898\,154\,535\,072 = 2^5 \cdot 3^2 \cdot 11 \cdot 13 \cdot 23 \cdot 29 \cdot 31 \cdot 53 \cdot 59 \cdot 61 \cdot 67 \cdot 71$$
$$1\,791\,242\,627\,540\,058\,576 = 2^4 \cdot 3^3 \cdot 13 \cdot 17 \cdot 23 \cdot 29 \cdot 31 \cdot 53 \cdot 59 \cdot 61 \cdot 67 \cdot 71$$
$$3\,894\,005\,712\,043\,605\,600 = 2^5 \cdot 3^3 \cdot 5^2 \cdot 13 \cdot 17 \cdot 29 \cdot 31 \cdot 53 \cdot 59 \cdot 61 \cdot 67 \cdot 71$$
$$7\,850\,261\,662\,089\,028\,100 = 2^2 \cdot 3^2 \cdot 5^2 \cdot 7^2 \cdot 13 \cdot 17 \cdot 29 \cdot 31 \cdot 53 \cdot 59 \cdot 61 \cdot 67 \cdot 71$$
$$15\,264\,502\,391\,210\,933\,952 = 2^6 \cdot 3^3 \cdot 7^2 \cdot 13 \cdot 17 \cdot 29 \cdot 31 \cdot 53 \cdot 59 \cdot 61 \cdot 67 \cdot 71$$
$$27\,593\,523\,553\,342\,842\,144 = 2^5 \cdot 3^3 \cdot 7^2 \cdot 17 \cdot 29 \cdot 31 \cdot 47 \cdot 53 \cdot 59 \cdot 61 \cdot 67 \cdot 71$$
$$47\,011\,188\,276\,065\,582\,912 = 2^6 \cdot 7^2 \cdot 17 \cdot 23 \cdot 29 \cdot 31 \cdot 47 \cdot 53 \cdot 59 \cdot 61 \cdot 67 \cdot 71$$
$$75\,553\,695\,443\,676\,829\,680 = 2^4 \cdot 3^2 \cdot 5 \cdot 7 \cdot 17 \cdot 23 \cdot 29 \cdot 31 \cdot 47 \cdot 53 \cdot 59 \cdot 61 \cdot 71$$
$$114\,633\,193\,086\,957\,948\,480 = 2^6 \cdot 3^2 \cdot 5 \cdot 7 \cdot 11 \cdot 17 \cdot 23 \cdot 31 \cdot 47 \cdot 53 \cdot 59 \cdot 61 \cdot 67 \cdot 71$$
$$164\,307\,576\,757\,973\,059\,488 = 2^5 \cdot 3 \cdot 7 \cdot 11 \cdot 17 \cdot 23 \cdot 31 \cdot 43 \cdot 47 \cdot 53 \cdot 59 \cdot 61 \cdot 67 \cdot 71$$
$$222\,610\,265\,284\,995\,758\,016 = 2^6 \cdot 3^2 \cdot 7^2 \cdot 11 \cdot 17 \cdot 23 \cdot 43 \cdot 47 \cdot 53 \cdot 59 \cdot 61 \cdot 67 \cdot 71$$
$$285\,219\,402\,396\,400\,814\,958 = 2 \cdot 3^2 \cdot 7^2 \cdot 11 \cdot 17 \cdot 23 \cdot 41 \cdot 43 \cdot 47 \cdot 53 \cdot 59 \cdot 61 \cdot 67 \cdot 71$$
$$345\,720\,487\,753\,213\,109\,040 = 2^4 \cdot 3 \cdot 5 \cdot 7^2 \cdot 17 \cdot 23 \cdot 41 \cdot 43 \cdot 47 \cdot 53 \cdot 59 \cdot 61 \cdot 67 \cdot 71$$
$$396\,561\,735\,952\,215\,036\,840 = 2^3 \cdot 3^2 \cdot 5 \cdot 7^2 \cdot 13 \cdot 23 \cdot 41 \cdot 43 \cdot 47 \cdot 53 \cdot 59 \cdot 61 \cdot 67 \cdot 71$$
$$430\,552\,741\,890\,976\,325\,712 = 2^4 \cdot 3^2 \cdot 7 \cdot 13 \cdot 19 \cdot 23 \cdot 41 \cdot 43 \cdot 47 \cdot 53 \cdot 59 \cdot 61 \cdot 67 \cdot 71$$
$$442\,512\,540\,276\,836\,779\,204 = 2^2 \cdot 7 \cdot 13 \cdot 19 \cdot 23 \cdot 37 \cdot 41 \cdot 43 \cdot 47 \cdot 53 \cdot 59 \cdot 61 \cdot 67 \cdot 71$$

<u>Row 73</u>

$$73 \text{ is Prime}$$

$$2\,628 = 2^2 \cdot 3^2 \cdot 73$$

$$62\,196 = 2^2 \cdot 3 \cdot 71 \cdot 73$$

$$1\,088\,430 = 2 \cdot 3 \cdot 5 \cdot 7 \cdot 71 \cdot 73$$

$$15\,020\,334 = 2 \cdot 3^2 \cdot 7 \cdot 23 \cdot 71 \cdot 73$$

$$170\,230\,452 = 2^2 \cdot 3 \cdot 7 \cdot 17 \cdot 23 \cdot 71 \cdot 73$$

$$1\,629\,348\,612 = 2^2 \cdot 3 \cdot 17 \cdot 23 \cdot 67 \cdot 71 \cdot 73$$

$$13\,442\,126\,049 = 3^2 \cdot 11 \cdot 17 \cdot 23 \cdot 67 \cdot 71 \cdot 73$$

$$97\,082\,021\,465 = 5 \cdot 11 \cdot 13 \cdot 17 \cdot 23 \cdot 67 \cdot 71 \cdot 73$$

$$621\,324\,937\,376 = 2^5 \cdot 11 \cdot 13 \cdot 17 \cdot 23 \cdot 67 \cdot 71 \cdot 73$$

$$3\,558\,497\,368\,608 = 2^5 \cdot 3^2 \cdot 7 \cdot 13 \cdot 17 \cdot 23 \cdot 67 \cdot 71 \cdot 73$$

$$18\,385\,569\,737\,808 = 2^4 \cdot 3 \cdot 7 \cdot 13 \cdot 17 \cdot 23 \cdot 31 \cdot 67 \cdot 71 \cdot 73$$

$$86\,270\,750\,308\,176 = 2^4 \cdot 3 \cdot 7 \cdot 17 \cdot 23 \cdot 31 \cdot 61 \cdot 67 \cdot 71 \cdot 73$$

$$369\,731\,787\,035\,040 = 2^5 \cdot 3^2 \cdot 5 \cdot 17 \cdot 23 \cdot 31 \cdot 61 \cdot 67 \cdot 71 \cdot 73$$

$$1\,454\,278\,362\,337\,824 = 2^5 \cdot 3 \cdot 17 \cdot 23 \cdot 31 \cdot 59 \cdot 61 \cdot 67 \cdot 71 \cdot 73$$

$$5\,271\,759\,063\,474\,612 = 2^2 \cdot 3 \cdot 17 \cdot 23 \cdot 29 \cdot 31 \cdot 59 \cdot 61 \cdot 67 \cdot 71 \cdot 73$$

$$17\,675\,898\,036\,356\,052 = 2^2 \cdot 3^2 \cdot 19 \cdot 23 \cdot 29 \cdot 31 \cdot 59 \cdot 61 \cdot 67 \cdot 71 \cdot 73$$

$$54\,991\,682\,779\,774\,384 = 2^4 \cdot 7 \cdot 19 \cdot 23 \cdot 29 \cdot 31 \cdot 59 \cdot 61 \cdot 67 \cdot 71 \cdot 73$$

$$159\,186\,450\,151\,978\,480 = 2^4 \cdot 5 \cdot 7 \cdot 11 \cdot 23 \cdot 29 \cdot 31 \cdot 59 \cdot 61 \cdot 67 \cdot 71 \cdot 73$$

$$429\,803\,415\,410\,341\,896 = 2^3 \cdot 3^3 \cdot 7 \cdot 11 \cdot 23 \cdot 29 \cdot 31 \cdot 59 \cdot 61 \cdot 67 \cdot 71 \cdot 73$$

$$1\,084\,741\,953\,178\,481\,928 = 2^3 \cdot 3^2 \cdot 11 \cdot 23 \cdot 29 \cdot 31 \cdot 53 \cdot 59 \cdot 61 \cdot 67 \cdot 71 \cdot 73$$

$$2\,563\,935\,525\,694\,593\,648 = 2^4 \cdot 3^2 \cdot 13 \cdot 23 \cdot 29 \cdot 31 \cdot 53 \cdot 59 \cdot 61 \cdot 67 \cdot 71 \cdot 73$$

$$5\,685\,248\,339\,583\,664\,176 = 2^4 \cdot 3^3 \cdot 13 \cdot 17 \cdot 29 \cdot 31 \cdot 53 \cdot 59 \cdot 61 \cdot 67 \cdot 71 \cdot 73$$

$$11\,844\,267\,374\,132\,633\,700 = 2^2 \cdot 3^2 \cdot 5^2 \cdot 13 \cdot 17 \cdot 29 \cdot 31 \cdot 53 \cdot 59 \cdot 61 \cdot 67 \cdot 71 \cdot 73$$

$$23\,214\,764\,053\,299\,962\,052 = 2^2 \cdot 3^2 \cdot 7^2 \cdot 13 \cdot 17 \cdot 29 \cdot 31 \cdot 53 \cdot 59 \cdot 61 \cdot 67 \cdot 71 \cdot 73$$

$$42\,858\,025\,944\,553\,776\,096 = 2^5 \cdot 3^3 \cdot 7^2 \cdot 17 \cdot 29 \cdot 31 \cdot 53 \cdot 59 \cdot 61 \cdot 67 \cdot 71 \cdot 73$$

$$74\,604\,711\,829\,408\,425\,056 = 2^5 \cdot 7^2 \cdot 17 \cdot 29 \cdot 31 \cdot 47 \cdot 53 \cdot 59 \cdot 61 \cdot 67 \cdot 71 \cdot 73$$

$$122\,564\,883\,719\,742\,412\,592 = 2^4 \cdot 7 \cdot 17 \cdot 23 \cdot 29 \cdot 31 \cdot 47 \cdot 53 \cdot 59 \cdot 61 \cdot 67 \cdot 71 \cdot 73$$

$$190\,186\,888\,530\,634\,778\,160 = 2^4 \cdot 3^2 \cdot 5 \cdot 7 \cdot 17 \cdot 23 \cdot 31 \cdot 47 \cdot 53 \cdot 59 \cdot 61 \cdot 67 \cdot 71 \cdot 73$$

$$278\,940\,769\,844\,931\,007\,968 = 2^5 \cdot 3 \cdot 7 \cdot 11 \cdot 17 \cdot 23 \cdot 31 \cdot 47 \cdot 53 \cdot 59 \cdot 61 \cdot 67 \cdot 71 \cdot 73$$

$$386\,917\,842\,042\,968\,817\,504 = 2^5 \cdot 3 \cdot 7 \cdot 11 \cdot 17 \cdot 23 \cdot 43 \cdot 47 \cdot 53 \cdot 59 \cdot 61 \cdot 67 \cdot 71 \cdot 73$$

$$507\,829\,667\,681\,396\,572\,974 = 2 \cdot 3^2 \cdot 7^2 \cdot 11 \cdot 17 \cdot 23 \cdot 43 \cdot 47 \cdot 53 \cdot 59 \cdot 61 \cdot 67 \cdot 71 \cdot 73$$

$$630\,939\,890\,149\,613\,923\,998 = 2 \cdot 3 \cdot 7^2 \cdot 17 \cdot 23 \cdot 41 \cdot 43 \cdot 47 \cdot 53 \cdot 59 \cdot 61 \cdot 67 \cdot 71 \cdot 73$$

$$742\,282\,223\,705\,428\,145\,880 = 2^3 \cdot 3 \cdot 5 \cdot 7^2 \cdot 23 \cdot 41 \cdot 43 \cdot 47 \cdot 53 \cdot 59 \cdot 61 \cdot 67 \cdot 71 \cdot 73$$

$$827\,114\,477\,843\,191\,362\,552 = 2^3 \cdot 3^2 \cdot 7 \cdot 13 \cdot 23 \cdot 41 \cdot 43 \cdot 47 \cdot 53 \cdot 59 \cdot 61 \cdot 67 \cdot 71 \cdot 73$$

$$873\,065\,282\,167\,813\,104\,916 = 2^2 \cdot 7 \cdot 13 \cdot 19 \cdot 23 \cdot 41 \cdot 43 \cdot 47 \cdot 53 \cdot 59 \cdot 61 \cdot 67 \cdot 71 \cdot 73$$

<u>Row 74</u>

$$74 = 2 \cdot 37$$
$$2\,701 = 37 \cdot 73$$
$$64\,824 = 2^3 \cdot 3 \cdot 37 \cdot 73$$
$$1\,150\,626 = 2 \cdot 3 \cdot 37 \cdot 71 \cdot 73$$
$$16\,108\,764 = 2^2 \cdot 3 \cdot 7 \cdot 37 \cdot 71 \cdot 73$$
$$185\,250\,786 = 2 \cdot 3 \cdot 7 \cdot 23 \cdot 37 \cdot 71 \cdot 73$$
$$1\,700\,570\,064 = 2^3 \cdot 3 \cdot 17 \cdot 23 \cdot 37 \cdot 71 \cdot 73$$
$$15\,071\,474\,661 = 3 \cdot 17 \cdot 23 \cdot 37 \cdot 67 \cdot 71 \cdot 73$$
$$110\,524\,147\,514 = 2 \cdot 11 \cdot 17 \cdot 23 \cdot 37 \cdot 67 \cdot 71 \cdot 73$$
$$718\,406\,958\,841 = 11 \cdot 13 \cdot 17 \cdot 23 \cdot 37 \cdot 67 \cdot 71 \cdot 73$$
$$4\,179\,822\,305\,984 = 2^6 \cdot 13 \cdot 17 \cdot 23 \cdot 37 \cdot 67 \cdot 71 \cdot 73$$
$$21\,944\,067\,106\,416 = 2^4 \cdot 3 \cdot 7 \cdot 13 \cdot 17 \cdot 23 \cdot 37 \cdot 67 \cdot 71 \cdot 73$$
$$104\,656\,320\,045\,984 = 2^5 \cdot 3 \cdot 7 \cdot 17 \cdot 23 \cdot 31 \cdot 37 \cdot 67 \cdot 71 \cdot 73$$
$$456\,002\,537\,343\,216 = 2^4 \cdot 3 \cdot 17 \cdot 23 \cdot 31 \cdot 37 \cdot 61 \cdot 67 \cdot 71 \cdot 73$$
$$1\,824\,010\,149\,372\,864 = 2^6 \cdot 3 \cdot 17 \cdot 23 \cdot 31 \cdot 37 \cdot 61 \cdot 67 \cdot 71 \cdot 73$$
$$6\,726\,037\,425\,812\,436 = 2^2 \cdot 3 \cdot 17 \cdot 23 \cdot 31 \cdot 37 \cdot 59 \cdot 61 \cdot 67 \cdot 71 \cdot 73$$
$$22\,947\,657\,099\,830\,664 = 2^3 \cdot 3 \cdot 23 \cdot 29 \cdot 31 \cdot 37 \cdot 59 \cdot 61 \cdot 67 \cdot 71 \cdot 73$$
$$72\,667\,580\,816\,130\,436 = 2^2 \cdot 19 \cdot 23 \cdot 29 \cdot 31 \cdot 37 \cdot 59 \cdot 61 \cdot 67 \cdot 71 \cdot 73$$
$$214\,178\,132\,931\,752\,864 = 2^5 \cdot 7 \cdot 23 \cdot 29 \cdot 31 \cdot 37 \cdot 59 \cdot 61 \cdot 67 \cdot 71 \cdot 73$$
$$588\,989\,865\,562\,320\,376 = 2^3 \cdot 7 \cdot 11 \cdot 23 \cdot 29 \cdot 31 \cdot 37 \cdot 59 \cdot 61 \cdot 67 \cdot 71 \cdot 73$$
$$1\,514\,545\,368\,588\,823\,824 = 2^4 \cdot 3^2 \cdot 11 \cdot 23 \cdot 29 \cdot 31 \cdot 37 \cdot 59 \cdot 61 \cdot 67 \cdot 71 \cdot 73$$
$$3\,648\,677\,478\,873\,075\,576 = 2^3 \cdot 3^2 \cdot 23 \cdot 29 \cdot 31 \cdot 37 \cdot 53 \cdot 59 \cdot 61 \cdot 67 \cdot 71 \cdot 73$$
$$8\,249\,183\,865\,278\,257\,824 = 2^5 \cdot 3^2 \cdot 13 \cdot 29 \cdot 31 \cdot 37 \cdot 53 \cdot 59 \cdot 61 \cdot 67 \cdot 71 \cdot 73$$
$$17\,529\,515\,713\,716\,297\,876 = 2^2 \cdot 3^2 \cdot 13 \cdot 17 \cdot 29 \cdot 31 \cdot 37 \cdot 53 \cdot 59 \cdot 61 \cdot 67 \cdot 71 \cdot 73$$
$$35\,059\,031\,427\,432\,595\,752 = 2^3 \cdot 3^2 \cdot 13 \cdot 17 \cdot 29 \cdot 31 \cdot 37 \cdot 53 \cdot 59 \cdot 61 \cdot 67 \cdot 71 \cdot 73$$
$$66\,072\,789\,997\,853\,738\,148 = 2^2 \cdot 3^2 \cdot 7^2 \cdot 17 \cdot 29 \cdot 31 \cdot 37 \cdot 53 \cdot 59 \cdot 61 \cdot 67 \cdot 71 \cdot 73$$
$$117\,462\,737\,773\,962\,201\,152 = 2^6 \cdot 7^2 \cdot 17 \cdot 29 \cdot 31 \cdot 37 \cdot 53 \cdot 59 \cdot 61 \cdot 67 \cdot 71 \cdot 73$$
$$197\,169\,595\,549\,150\,837\,648 = 2^4 \cdot 7 \cdot 17 \cdot 29 \cdot 31 \cdot 37 \cdot 47 \cdot 53 \cdot 59 \cdot 61 \cdot 67 \cdot 71 \cdot 73$$
$$312\,751\,772\,250\,377\,190\,752 = 2^5 \cdot 7 \cdot 17 \cdot 23 \cdot 31 \cdot 37 \cdot 47 \cdot 53 \cdot 59 \cdot 61 \cdot 67 \cdot 71 \cdot 73$$
$$469\,127\,658\,375\,565\,786\,128 = 2^4 \cdot 3 \cdot 7 \cdot 17 \cdot 23 \cdot 31 \cdot 37 \cdot 47 \cdot 53 \cdot 59 \cdot 61 \cdot 67 \cdot 71 \cdot 73$$
$$665\,858\,611\,887\,899\,825\,472 = 2^6 \cdot 3 \cdot 7 \cdot 11 \cdot 17 \cdot 23 \cdot 37 \cdot 47 \cdot 53 \cdot 59 \cdot 61 \cdot 67 \cdot 71 \cdot 73$$
$$894\,747\,509\,724\,365\,390\,478 = 2 \cdot 3 \cdot 7 \cdot 11 \cdot 17 \cdot 23 \cdot 37 \cdot 43 \cdot 47 \cdot 53 \cdot 59 \cdot 61 \cdot 67 \cdot 71 \cdot 73$$
$$1\,138\,769\,557\,831\,010\,496\,972 = 2^2 \cdot 3 \cdot 7^2 \cdot 17 \cdot 23 \cdot 37 \cdot 43 \cdot 47 \cdot 53 \cdot 59 \cdot 61 \cdot 67 \cdot 71 \cdot 73$$
$$1\,373\,222\,113\,855\,042\,069\,878 = 2 \cdot 3 \cdot 7^2 \cdot 23 \cdot 37 \cdot 41 \cdot 43 \cdot 47 \cdot 53 \cdot 59 \cdot 61 \cdot 67 \cdot 71 \cdot 73$$
$$1\,569\,396\,701\,548\,619\,508\,432 = 2^4 \cdot 3 \cdot 7 \cdot 23 \cdot 37 \cdot 41 \cdot 43 \cdot 47 \cdot 53 \cdot 59 \cdot 61 \cdot 67 \cdot 71 \cdot 73$$
$$1\,700\,179\,760\,011\,004\,467\,468 = 2^2 \cdot 7 \cdot 13 \cdot 23 \cdot 37 \cdot 41 \cdot 43 \cdot 47 \cdot 53 \cdot 59 \cdot 61 \cdot 67 \cdot 71 \cdot 73$$
$$1\,746\,130\,564\,335\,626\,209\,832 = 2^3 \cdot 7 \cdot 13 \cdot 19 \cdot 23 \cdot 41 \cdot 43 \cdot 47 \cdot 53 \cdot 59 \cdot 61 \cdot 67 \cdot 71 \cdot 73$$

Row 75

$$75 = 3 \cdot 5^2$$
$$2\,775 = 3 \cdot 5^2 \cdot 37$$
$$67\,525 = 5^2 \cdot 37 \cdot 73$$
$$1\,215\,450 = 2 \cdot 3^2 \cdot 5^2 \cdot 37 \cdot 73$$
$$17\,259\,390 = 2 \cdot 3^2 \cdot 5 \cdot 37 \cdot 71 \cdot 73$$
$$201\,359\,550 = 2 \cdot 3 \cdot 5^2 \cdot 7 \cdot 37 \cdot 71 \cdot 73$$
$$1\,984\,829\,850 = 2 \cdot 3^2 \cdot 5^2 \cdot 23 \cdot 37 \cdot 71 \cdot 73$$
$$16\,871\,053\,725 = 3^2 \cdot 5^2 \cdot 17 \cdot 23 \cdot 37 \cdot 71 \cdot 73$$
$$125\,595\,622\,175 = 5^2 \cdot 17 \cdot 23 \cdot 37 \cdot 67 \cdot 71 \cdot 73$$
$$828\,931\,106\,355 = 3 \cdot 5 \cdot 11 \cdot 17 \cdot 23 \cdot 37 \cdot 67 \cdot 71 \cdot 73$$
$$4\,898\,229\,264\,825 = 3 \cdot 5^2 \cdot 13 \cdot 17 \cdot 23 \cdot 37 \cdot 67 \cdot 71 \cdot 73$$
$$26\,123\,889\,412\,400 = 2^4 \cdot 5^2 \cdot 13 \cdot 17 \cdot 23 \cdot 37 \cdot 67 \cdot 71 \cdot 73$$
$$126\,600\,387\,152\,400 = 2^4 \cdot 3^2 \cdot 5^2 \cdot 7 \cdot 17 \cdot 23 \cdot 37 \cdot 67 \cdot 71 \cdot 73$$
$$560\,658\,857\,389\,200 = 2^4 \cdot 3^2 \cdot 5^2 \cdot 17 \cdot 23 \cdot 31 \cdot 37 \cdot 67 \cdot 71 \cdot 73$$
$$2\,280\,012\,686\,716\,080 = 2^4 \cdot 3 \cdot 5 \cdot 17 \cdot 23 \cdot 31 \cdot 37 \cdot 61 \cdot 67 \cdot 71 \cdot 73$$
$$8\,550\,047\,575\,185\,300 = 2^2 \cdot 3^2 \cdot 5^2 \cdot 17 \cdot 23 \cdot 31 \cdot 37 \cdot 61 \cdot 67 \cdot 71 \cdot 73$$
$$29\,673\,694\,525\,643\,100 = 2^2 \cdot 3^2 \cdot 5^2 \cdot 23 \cdot 31 \cdot 37 \cdot 59 \cdot 61 \cdot 67 \cdot 71 \cdot 73$$
$$95\,615\,237\,915\,961\,100 = 2^2 \cdot 5^2 \cdot 23 \cdot 29 \cdot 31 \cdot 37 \cdot 59 \cdot 61 \cdot 67 \cdot 71 \cdot 73$$
$$286\,845\,713\,747\,883\,300 = 2^2 \cdot 3 \cdot 5^2 \cdot 23 \cdot 29 \cdot 31 \cdot 37 \cdot 59 \cdot 61 \cdot 67 \cdot 71 \cdot 73$$
$$803\,167\,998\,494\,073\,240 = 2^3 \cdot 3 \cdot 5 \cdot 7 \cdot 23 \cdot 29 \cdot 31 \cdot 37 \cdot 59 \cdot 61 \cdot 67 \cdot 71 \cdot 73$$
$$2\,103\,535\,234\,151\,144\,200 = 2^3 \cdot 5^2 \cdot 11 \cdot 23 \cdot 29 \cdot 31 \cdot 37 \cdot 59 \cdot 61 \cdot 67 \cdot 71 \cdot 73$$
$$5\,163\,222\,847\,461\,899\,400 = 2^3 \cdot 3^3 \cdot 5^2 \cdot 23 \cdot 29 \cdot 31 \cdot 37 \cdot 59 \cdot 61 \cdot 67 \cdot 71 \cdot 73$$
$$11\,897\,861\,344\,151\,333\,400 = 2^3 \cdot 3^3 \cdot 5^2 \cdot 29 \cdot 31 \cdot 37 \cdot 53 \cdot 59 \cdot 61 \cdot 67 \cdot 71 \cdot 73$$
$$25\,778\,699\,578\,994\,555\,700 = 2^2 \cdot 3^2 \cdot 5^2 \cdot 13 \cdot 29 \cdot 31 \cdot 37 \cdot 53 \cdot 59 \cdot 61 \cdot 67 \cdot 71 \cdot 73$$
$$52\,588\,547\,141\,148\,893\,628 = 2^2 \cdot 3^3 \cdot 13 \cdot 17 \cdot 29 \cdot 31 \cdot 37 \cdot 53 \cdot 59 \cdot 61 \cdot 67 \cdot 71 \cdot 73$$
$$101\,131\,821\,425\,286\,333\,900 = 2^2 \cdot 3^3 \cdot 5^2 \cdot 17 \cdot 29 \cdot 31 \cdot 37 \cdot 53 \cdot 59 \cdot 61 \cdot 67 \cdot 71 \cdot 73$$
$$183\,535\,527\,771\,815\,939\,300 = 2^2 \cdot 5^2 \cdot 7^2 \cdot 17 \cdot 29 \cdot 31 \cdot 37 \cdot 53 \cdot 59 \cdot 61 \cdot 67 \cdot 71 \cdot 73$$
$$314\,632\,333\,323\,113\,038\,800 = 2^4 \cdot 3 \cdot 5^2 \cdot 7 \cdot 17 \cdot 29 \cdot 31 \cdot 37 \cdot 53 \cdot 59 \cdot 61 \cdot 67 \cdot 71 \cdot 73$$
$$509\,921\,367\,799\,528\,028\,400 = 2^4 \cdot 3 \cdot 5^2 \cdot 7 \cdot 17 \cdot 31 \cdot 37 \cdot 47 \cdot 53 \cdot 59 \cdot 61 \cdot 67 \cdot 71 \cdot 73$$
$$781\,879\,430\,625\,942\,976\,880 = 2^4 \cdot 5 \cdot 7 \cdot 17 \cdot 23 \cdot 31 \cdot 37 \cdot 47 \cdot 53 \cdot 59 \cdot 61 \cdot 67 \cdot 71 \cdot 73$$
$$1\,134\,986\,270\,263\,465\,611\,600 = 2^4 \cdot 3^2 \cdot 5^2 \cdot 7 \cdot 17 \cdot 23 \cdot 37 \cdot 47 \cdot 53 \cdot 59 \cdot 61 \cdot 67 \cdot 71 \cdot 73$$
$$1\,560\,606\,121\,612\,265\,215\,950 = 2 \cdot 3^2 \cdot 5^2 \cdot 7 \cdot 11 \cdot 17 \cdot 23 \cdot 37 \cdot 47 \cdot 53 \cdot 59 \cdot 61 \cdot 67 \cdot 71 \cdot 73$$
$$2\,033\,517\,067\,555\,375\,887\,450 = 2 \cdot 3 \cdot 5^2 \cdot 7 \cdot 17 \cdot 23 \cdot 37 \cdot 43 \cdot 47 \cdot 53 \cdot 59 \cdot 61 \cdot 67 \cdot 71 \cdot 73$$
$$2\,511\,991\,671\,686\,052\,566\,850 = 2 \cdot 3^2 \cdot 5^2 \cdot 7^2 \cdot 23 \cdot 37 \cdot 43 \cdot 47 \cdot 53 \cdot 59 \cdot 61 \cdot 67 \cdot 71 \cdot 73$$
$$2\,942\,618\,815\,403\,661\,578\,310 = 2 \cdot 3^2 \cdot 5 \cdot 7 \cdot 23 \cdot 37 \cdot 41 \cdot 43 \cdot 47 \cdot 53 \cdot 59 \cdot 61 \cdot 67 \cdot 71 \cdot 73$$
$$3\,269\,576\,461\,559\,623\,975\,900 = 2^2 \cdot 5^2 \cdot 7 \cdot 23 \cdot 37 \cdot 41 \cdot 43 \cdot 47 \cdot 53 \cdot 59 \cdot 61 \cdot 67 \cdot 71 \cdot 73$$
$$3\,446\,310\,324\,346\,630\,677\,300 = 2^2 \cdot 3 \cdot 5^2 \cdot 7 \cdot 13 \cdot 23 \cdot 41 \cdot 43 \cdot 47 \cdot 53 \cdot 59 \cdot 61 \cdot 67 \cdot 71 \cdot 73$$

Row 76

$$76 = 2^2 \cdot 19$$
$$2\,850 = 2 \cdot 3 \cdot 5^2 \cdot 19$$
$$70\,300 = 2^2 \cdot 5^2 \cdot 19 \cdot 37$$
$$1\,282\,975 = 5^2 \cdot 19 \cdot 37 \cdot 73$$
$$18\,474\,840 = 2^3 \cdot 3^2 \cdot 5 \cdot 19 \cdot 37 \cdot 73$$
$$218\,618\,940 = 2^2 \cdot 3 \cdot 5 \cdot 19 \cdot 37 \cdot 71 \cdot 73$$
$$2\,186\,189\,400 = 2^3 \cdot 3 \cdot 5^2 \cdot 19 \cdot 37 \cdot 71 \cdot 73$$
$$18\,855\,883\,575 = 3^2 \cdot 5^2 \cdot 19 \cdot 23 \cdot 37 \cdot 71 \cdot 73$$
$$142\,466\,675\,900 = 2^2 \cdot 5^2 \cdot 17 \cdot 19 \cdot 23 \cdot 37 \cdot 71 \cdot 73$$
$$954\,526\,728\,530 = 2 \cdot 5 \cdot 17 \cdot 19 \cdot 23 \cdot 37 \cdot 67 \cdot 71 \cdot 73$$
$$5\,727\,160\,371\,180 = 2^2 \cdot 3 \cdot 5 \cdot 17 \cdot 19 \cdot 23 \cdot 37 \cdot 67 \cdot 71 \cdot 73$$
$$31\,022\,118\,677\,225 = 5^2 \cdot 13 \cdot 17 \cdot 19 \cdot 23 \cdot 37 \cdot 67 \cdot 71 \cdot 73$$
$$152\,724\,276\,564\,800 = 2^6 \cdot 5^2 \cdot 17 \cdot 19 \cdot 23 \cdot 37 \cdot 67 \cdot 71 \cdot 73$$
$$687\,259\,244\,541\,600 = 2^5 \cdot 3^2 \cdot 5^2 \cdot 17 \cdot 19 \cdot 23 \cdot 37 \cdot 67 \cdot 71 \cdot 73$$
$$2\,840\,671\,544\,105\,280 = 2^6 \cdot 3 \cdot 5 \cdot 17 \cdot 19 \cdot 23 \cdot 31 \cdot 37 \cdot 67 \cdot 71 \cdot 73$$
$$10\,830\,060\,261\,901\,380 = 2^2 \cdot 3 \cdot 5 \cdot 17 \cdot 19 \cdot 23 \cdot 31 \cdot 37 \cdot 61 \cdot 67 \cdot 71 \cdot 73$$
$$38\,223\,742\,100\,828\,400 = 2^4 \cdot 3^2 \cdot 5^2 \cdot 19 \cdot 23 \cdot 31 \cdot 37 \cdot 61 \cdot 67 \cdot 71 \cdot 73$$
$$125\,288\,932\,441\,604\,200 = 2^3 \cdot 5^2 \cdot 19 \cdot 23 \cdot 31 \cdot 37 \cdot 59 \cdot 61 \cdot 67 \cdot 71 \cdot 73$$
$$382\,460\,951\,663\,844\,400 = 2^4 \cdot 5^2 \cdot 23 \cdot 29 \cdot 31 \cdot 37 \cdot 59 \cdot 61 \cdot 67 \cdot 71 \cdot 73$$
$$1\,090\,013\,712\,241\,956\,540 = 2^2 \cdot 3 \cdot 5 \cdot 19 \cdot 23 \cdot 29 \cdot 31 \cdot 37 \cdot 59 \cdot 61 \cdot 67 \cdot 71 \cdot 73$$
$$2\,906\,703\,232\,645\,217\,440 = 2^5 \cdot 5 \cdot 19 \cdot 23 \cdot 29 \cdot 31 \cdot 37 \cdot 59 \cdot 61 \cdot 67 \cdot 71 \cdot 73$$
$$7\,266\,758\,081\,613\,043\,600 = 2^4 \cdot 5^2 \cdot 19 \cdot 23 \cdot 29 \cdot 31 \cdot 37 \cdot 59 \cdot 61 \cdot 67 \cdot 71 \cdot 73$$
$$17\,061\,084\,191\,613\,232\,800 = 2^5 \cdot 3^3 \cdot 5^2 \cdot 19 \cdot 29 \cdot 31 \cdot 37 \cdot 59 \cdot 61 \cdot 67 \cdot 71 \cdot 73$$
$$37\,676\,560\,923\,145\,889\,100 = 2^2 \cdot 3^2 \cdot 5^2 \cdot 19 \cdot 29 \cdot 31 \cdot 37 \cdot 53 \cdot 59 \cdot 61 \cdot 67 \cdot 71 \cdot 73$$
$$78\,367\,246\,720\,143\,449\,328 = 2^4 \cdot 3^2 \cdot 13 \cdot 19 \cdot 29 \cdot 31 \cdot 37 \cdot 53 \cdot 59 \cdot 61 \cdot 67 \cdot 71 \cdot 73$$
$$153\,720\,368\,566\,435\,227\,528 = 2^3 \cdot 3^3 \cdot 17 \cdot 19 \cdot 29 \cdot 31 \cdot 37 \cdot 53 \cdot 59 \cdot 61 \cdot 67 \cdot 71 \cdot 73$$
$$284\,667\,349\,197\,102\,273\,200 = 2^4 \cdot 5^2 \cdot 17 \cdot 19 \cdot 29 \cdot 31 \cdot 37 \cdot 53 \cdot 59 \cdot 61 \cdot 67 \cdot 71 \cdot 73$$
$$498\,167\,861\,094\,928\,978\,100 = 2^2 \cdot 5^2 \cdot 7 \cdot 17 \cdot 19 \cdot 29 \cdot 31 \cdot 37 \cdot 53 \cdot 59 \cdot 61 \cdot 67 \cdot 71 \cdot 73$$
$$824\,553\,701\,122\,641\,067\,200 = 2^6 \cdot 3 \cdot 5^2 \cdot 7 \cdot 17 \cdot 19 \cdot 31 \cdot 37 \cdot 53 \cdot 59 \cdot 61 \cdot 67 \cdot 71 \cdot 73$$
$$1\,291\,800\,798\,425\,471\,005\,280 = 2^5 \cdot 5 \cdot 7 \cdot 17 \cdot 19 \cdot 31 \cdot 37 \cdot 47 \cdot 53 \cdot 59 \cdot 61 \cdot 67 \cdot 71 \cdot 73$$
$$1\,916\,865\,700\,889\,408\,588\,480 = 2^6 \cdot 5 \cdot 7 \cdot 17 \cdot 19 \cdot 23 \cdot 37 \cdot 47 \cdot 53 \cdot 59 \cdot 61 \cdot 67 \cdot 71 \cdot 73$$
$$2\,695\,592\,391\,875\,730\,827\,550 = 2 \cdot 3^2 \cdot 5^2 \cdot 7 \cdot 17 \cdot 19 \cdot 23 \cdot 37 \cdot 47 \cdot 53 \cdot 59 \cdot 61 \cdot 67 \cdot 71 \cdot 73$$
$$3\,594\,123\,189\,167\,641\,103\,400 = 2^3 \cdot 3 \cdot 5^2 \cdot 7 \cdot 17 \cdot 19 \cdot 23 \cdot 37 \cdot 47 \cdot 53 \cdot 59 \cdot 61 \cdot 67 \cdot 71 \cdot 73$$
$$4\,545\,508\,739\,241\,428\,454\,300 = 2^2 \cdot 3 \cdot 5^2 \cdot 7 \cdot 19 \cdot 23 \cdot 37 \cdot 43 \cdot 47 \cdot 53 \cdot 59 \cdot 61 \cdot 67 \cdot 71 \cdot 73$$
$$5\,454\,610\,487\,089\,714\,145\,160 = 2^3 \cdot 3^2 \cdot 5 \cdot 7 \cdot 19 \cdot 23 \cdot 37 \cdot 43 \cdot 47 \cdot 53 \cdot 59 \cdot 61 \cdot 67 \cdot 71 \cdot 73$$
$$6\,212\,195\,276\,963\,285\,554\,210 = 2 \cdot 5 \cdot 7 \cdot 19 \cdot 23 \cdot 37 \cdot 41 \cdot 43 \cdot 47 \cdot 53 \cdot 59 \cdot 61 \cdot 67 \cdot 71 \cdot 73$$
$$6\,715\,886\,785\,906\,254\,653\,200 = 2^4 \cdot 5^2 \cdot 7 \cdot 19 \cdot 23 \cdot 41 \cdot 43 \cdot 47 \cdot 53 \cdot 59 \cdot 61 \cdot 67 \cdot 71 \cdot 73$$
$$6\,892\,620\,648\,693\,261\,354\,600 = 2^3 \cdot 3 \cdot 5^2 \cdot 7 \cdot 13 \cdot 23 \cdot 41 \cdot 43 \cdot 47 \cdot 53 \cdot 59 \cdot 61 \cdot 67 \cdot 71 \cdot 73$$

Row 77

$$77 = 7 \cdot 11$$
$$2\,926 = 2 \cdot 7 \cdot 11 \cdot 19$$
$$73\,150 = 2 \cdot 5^2 \cdot 7 \cdot 11 \cdot 19$$
$$1\,353\,275 = 5^2 \cdot 7 \cdot 11 \cdot 19 \cdot 37$$
$$19\,757\,815 = 5 \cdot 7 \cdot 11 \cdot 19 \cdot 37 \cdot 73$$
$$237\,093\,780 = 2^2 \cdot 3 \cdot 5 \cdot 7 \cdot 11 \cdot 19 \cdot 37 \cdot 73$$
$$2\,404\,808\,340 = 2^2 \cdot 3 \cdot 5 \cdot 11 \cdot 19 \cdot 37 \cdot 71 \cdot 73$$
$$21\,042\,072\,975 = 3 \cdot 5^2 \cdot 7 \cdot 11 \cdot 19 \cdot 37 \cdot 71 \cdot 73$$
$$161\,322\,559\,475 = 5^2 \cdot 7 \cdot 11 \cdot 19 \cdot 23 \cdot 37 \cdot 71 \cdot 73$$
$$1\,096\,993\,404\,430 = 2 \cdot 5 \cdot 7 \cdot 11 \cdot 17 \cdot 19 \cdot 23 \cdot 37 \cdot 71 \cdot 73$$
$$6\,681\,687\,099\,710 = 2 \cdot 5 \cdot 7 \cdot 17 \cdot 19 \cdot 23 \cdot 37 \cdot 67 \cdot 71 \cdot 73$$
$$36\,749\,279\,048\,405 = 5 \cdot 7 \cdot 11 \cdot 17 \cdot 19 \cdot 23 \cdot 37 \cdot 67 \cdot 71 \cdot 73$$
$$183\,746\,395\,242\,025 = 5^2 \cdot 7 \cdot 11 \cdot 17 \cdot 19 \cdot 23 \cdot 37 \cdot 67 \cdot 71 \cdot 73$$
$$839\,983\,521\,106\,400 = 2^5 \cdot 5^2 \cdot 11 \cdot 17 \cdot 19 \cdot 23 \cdot 37 \cdot 67 \cdot 71 \cdot 73$$
$$3\,527\,930\,788\,646\,880 = 2^5 \cdot 3 \cdot 5 \cdot 7 \cdot 11 \cdot 17 \cdot 19 \cdot 23 \cdot 37 \cdot 67 \cdot 71 \cdot 73$$
$$13\,670\,731\,806\,006\,660 = 2^2 \cdot 3 \cdot 5 \cdot 7 \cdot 11 \cdot 17 \cdot 19 \cdot 23 \cdot 31 \cdot 37 \cdot 67 \cdot 71 \cdot 73$$
$$49\,053\,802\,362\,729\,780 = 2^2 \cdot 3 \cdot 5 \cdot 7 \cdot 11 \cdot 19 \cdot 23 \cdot 31 \cdot 37 \cdot 61 \cdot 67 \cdot 71 \cdot 73$$
$$163\,512\,674\,542\,432\,600 = 2^3 \cdot 5^2 \cdot 7 \cdot 11 \cdot 19 \cdot 23 \cdot 31 \cdot 37 \cdot 61 \cdot 67 \cdot 71 \cdot 73$$
$$507\,749\,884\,105\,448\,600 = 2^3 \cdot 5^2 \cdot 7 \cdot 11 \cdot 23 \cdot 31 \cdot 37 \cdot 59 \cdot 61 \cdot 67 \cdot 71 \cdot 73$$
$$1\,472\,474\,663\,905\,800\,940 = 2^2 \cdot 5 \cdot 7 \cdot 11 \cdot 23 \cdot 29 \cdot 31 \cdot 37 \cdot 59 \cdot 61 \cdot 67 \cdot 71 \cdot 73$$
$$3\,996\,716\,944\,887\,173\,980 = 2^2 \cdot 5 \cdot 11 \cdot 19 \cdot 23 \cdot 29 \cdot 31 \cdot 37 \cdot 59 \cdot 61 \cdot 67 \cdot 71 \cdot 73$$
$$10\,173\,461\,314\,258\,261\,040 = 2^4 \cdot 5 \cdot 7 \cdot 19 \cdot 23 \cdot 29 \cdot 31 \cdot 37 \cdot 59 \cdot 61 \cdot 67 \cdot 71 \cdot 73$$
$$24\,327\,842\,273\,226\,276\,400 = 2^4 \cdot 5^2 \cdot 7 \cdot 11 \cdot 19 \cdot 29 \cdot 31 \cdot 37 \cdot 59 \cdot 61 \cdot 67 \cdot 71 \cdot 73$$
$$54\,737\,645\,114\,759\,121\,900 = 2^2 \cdot 3^2 \cdot 5^2 \cdot 7 \cdot 11 \cdot 19 \cdot 29 \cdot 31 \cdot 37 \cdot 59 \cdot 61 \cdot 67 \cdot 71 \cdot 73$$
$$116\,043\,807\,643\,289\,338\,428 = 2^2 \cdot 3^2 \cdot 7 \cdot 11 \cdot 19 \cdot 29 \cdot 31 \cdot 37 \cdot 53 \cdot 59 \cdot 61 \cdot 67 \cdot 71 \cdot 73$$
$$232\,087\,615\,286\,578\,676\,856 = 2^3 \cdot 3^2 \cdot 7 \cdot 11 \cdot 19 \cdot 29 \cdot 31 \cdot 37 \cdot 53 \cdot 59 \cdot 61 \cdot 67 \cdot 71 \cdot 73$$
$$438\,387\,717\,763\,537\,500\,728 = 2^3 \cdot 7 \cdot 11 \cdot 17 \cdot 19 \cdot 29 \cdot 31 \cdot 37 \cdot 53 \cdot 59 \cdot 61 \cdot 67 \cdot 71 \cdot 73$$
$$782\,835\,210\,292\,031\,251\,300 = 2^2 \cdot 5^2 \cdot 11 \cdot 17 \cdot 19 \cdot 29 \cdot 31 \cdot 37 \cdot 53 \cdot 59 \cdot 61 \cdot 67 \cdot 71 \cdot 73$$
$$1\,322\,721\,562\,217\,570\,045\,300 = 2^2 \cdot 5^2 \cdot 7^2 \cdot 11 \cdot 17 \cdot 19 \cdot 31 \cdot 37 \cdot 53 \cdot 59 \cdot 61 \cdot 67 \cdot 71 \cdot 73$$
$$2\,116\,354\,499\,548\,112\,072\,480 = 2^5 \cdot 5 \cdot 7^2 \cdot 11 \cdot 17 \cdot 19 \cdot 31 \cdot 37 \cdot 53 \cdot 59 \cdot 61 \cdot 67 \cdot 71 \cdot 73$$
$$3\,208\,666\,499\,314\,879\,593\,760 = 2^5 \cdot 5 \cdot 7^2 \cdot 11 \cdot 17 \cdot 19 \cdot 37 \cdot 47 \cdot 53 \cdot 59 \cdot 61 \cdot 67 \cdot 71 \cdot 73$$
$$4\,612\,458\,092\,765\,139\,416\,030 = 2 \cdot 5 \cdot 7^2 \cdot 11 \cdot 17 \cdot 19 \cdot 23 \cdot 37 \cdot 47 \cdot 53 \cdot 59 \cdot 61 \cdot 67 \cdot 71 \cdot 73$$
$$6\,289\,715\,581\,043\,371\,930\,950 = 2 \cdot 3 \cdot 5^2 \cdot 7^2 \cdot 17 \cdot 19 \cdot 23 \cdot 37 \cdot 47 \cdot 53 \cdot 59 \cdot 61 \cdot 67 \cdot 71 \cdot 73$$
$$8\,139\,631\,928\,409\,069\,557\,700 = 2^2 \cdot 3 \cdot 5^2 \cdot 7^2 \cdot 11 \cdot 19 \cdot 23 \cdot 37 \cdot 47 \cdot 53 \cdot 59 \cdot 61 \cdot 67 \cdot 71 \cdot 73$$
$$10\,000\,119\,226\,331\,142\,599\,460 = 2^2 \cdot 3 \cdot 5 \cdot 7 \cdot 11 \cdot 19 \cdot 23 \cdot 37 \cdot 43 \cdot 47 \cdot 53 \cdot 59 \cdot 61 \cdot 67 \cdot 71 \cdot 73$$
$$11\,666\,805\,764\,052\,999\,699\,370 = 2 \cdot 5 \cdot 7^2 \cdot 11 \cdot 19 \cdot 23 \cdot 37 \cdot 43 \cdot 47 \cdot 53 \cdot 59 \cdot 61 \cdot 67 \cdot 71 \cdot 73$$
$$12\,928\,082\,062\,869\,540\,207\,410 = 2 \cdot 5 \cdot 7^2 \cdot 11 \cdot 19 \cdot 23 \cdot 41 \cdot 43 \cdot 47 \cdot 53 \cdot 59 \cdot 61 \cdot 67 \cdot 71 \cdot 73$$
$$13\,608\,507\,434\,599\,516\,007\,800 = 2^3 \cdot 5^2 \cdot 7^2 \cdot 11 \cdot 23 \cdot 41 \cdot 43 \cdot 47 \cdot 53 \cdot 59 \cdot 61 \cdot 67 \cdot 71 \cdot 73$$

Pascal's Triangle — Prime Factorization — To Center Number (omitting 1's)

Row 78

$$78 = 2 \cdot 3 \cdot 13$$

$$3\,003 = 3 \cdot 7 \cdot 11 \cdot 13$$

$$76\,076 = 2^2 \cdot 7 \cdot 11 \cdot 13 \cdot 19$$

$$1\,426\,425 = 3 \cdot 5^2 \cdot 7 \cdot 11 \cdot 13 \cdot 19$$

$$21\,111\,090 = 2 \cdot 3 \cdot 5 \cdot 7 \cdot 11 \cdot 13 \cdot 19 \cdot 37$$

$$256\,851\,595 = 5 \cdot 7 \cdot 11 \cdot 13 \cdot 19 \cdot 37 \cdot 73$$

$$2\,641\,902\,120 = 2^3 \cdot 3^2 \cdot 5 \cdot 11 \cdot 13 \cdot 19 \cdot 37 \cdot 73$$

$$23\,446\,881\,315 = 3^2 \cdot 5 \cdot 11 \cdot 13 \cdot 19 \cdot 37 \cdot 71 \cdot 73$$

$$182\,364\,632\,450 = 2 \cdot 5^2 \cdot 7 \cdot 11 \cdot 13 \cdot 19 \cdot 37 \cdot 71 \cdot 73$$

$$1\,258\,315\,963\,905 = 3 \cdot 5 \cdot 7 \cdot 11 \cdot 13 \cdot 19 \cdot 23 \cdot 37 \cdot 71 \cdot 73$$

$$7\,778\,680\,504\,140 = 2^2 \cdot 3 \cdot 5 \cdot 7 \cdot 13 \cdot 17 \cdot 19 \cdot 23 \cdot 37 \cdot 71 \cdot 73$$

$$43\,430\,966\,148\,115 = 5 \cdot 7 \cdot 13 \cdot 17 \cdot 19 \cdot 23 \cdot 37 \cdot 67 \cdot 71 \cdot 73$$

$$220\,495\,674\,290\,430 = 2 \cdot 3 \cdot 5 \cdot 7 \cdot 11 \cdot 17 \cdot 19 \cdot 23 \cdot 37 \cdot 67 \cdot 71 \cdot 73$$

$$1\,023\,729\,916\,348\,425 = 3 \cdot 5^2 \cdot 11 \cdot 13 \cdot 17 \cdot 19 \cdot 23 \cdot 37 \cdot 67 \cdot 71 \cdot 73$$

$$4\,367\,914\,309\,753\,280 = 2^6 \cdot 5 \cdot 11 \cdot 13 \cdot 17 \cdot 19 \cdot 23 \cdot 37 \cdot 67 \cdot 71 \cdot 73$$

$$17\,198\,662\,594\,653\,540 = 2^2 \cdot 3^2 \cdot 5 \cdot 7 \cdot 11 \cdot 13 \cdot 17 \cdot 19 \cdot 23 \cdot 37 \cdot 67 \cdot 71 \cdot 73$$

$$62\,724\,534\,168\,736\,440 = 2^3 \cdot 3^2 \cdot 5 \cdot 7 \cdot 11 \cdot 13 \cdot 19 \cdot 23 \cdot 31 \cdot 37 \cdot 67 \cdot 71 \cdot 73$$

$$212\,566\,476\,905\,162\,380 = 2^2 \cdot 5 \cdot 7 \cdot 11 \cdot 13 \cdot 19 \cdot 23 \cdot 31 \cdot 37 \cdot 61 \cdot 67 \cdot 71 \cdot 73$$

$$671\,262\,558\,647\,881\,200 = 2^4 \cdot 3 \cdot 5^2 \cdot 7 \cdot 11 \cdot 13 \cdot 23 \cdot 31 \cdot 37 \cdot 61 \cdot 67 \cdot 71 \cdot 73$$

$$1\,980\,224\,548\,011\,249\,540 = 2^2 \cdot 3 \cdot 5 \cdot 7 \cdot 11 \cdot 13 \cdot 23 \cdot 31 \cdot 37 \cdot 59 \cdot 61 \cdot 67 \cdot 71 \cdot 73$$

$$5\,469\,191\,608\,792\,974\,920 = 2^3 \cdot 5 \cdot 11 \cdot 13 \cdot 23 \cdot 29 \cdot 31 \cdot 37 \cdot 59 \cdot 61 \cdot 67 \cdot 71 \cdot 73$$

$$14\,170\,178\,259\,145\,435\,020 = 2^2 \cdot 3 \cdot 5 \cdot 13 \cdot 19 \cdot 23 \cdot 29 \cdot 31 \cdot 37 \cdot 59 \cdot 61 \cdot 67 \cdot 71 \cdot 73$$

$$34\,501\,303\,587\,484\,537\,440 = 2^5 \cdot 3 \cdot 5 \cdot 7 \cdot 13 \cdot 19 \cdot 29 \cdot 31 \cdot 37 \cdot 59 \cdot 61 \cdot 67 \cdot 71 \cdot 73$$

$$79\,065\,487\,387\,985\,398\,300 = 2^2 \cdot 5^2 \cdot 7 \cdot 11 \cdot 13 \cdot 19 \cdot 29 \cdot 31 \cdot 37 \cdot 59 \cdot 61 \cdot 67 \cdot 71 \cdot 73$$

$$170\,781\,452\,758\,048\,460\,328 = 2^3 \cdot 3^3 \cdot 7 \cdot 11 \cdot 13 \cdot 19 \cdot 29 \cdot 31 \cdot 37 \cdot 59 \cdot 61 \cdot 67 \cdot 71 \cdot 73$$

$$348\,131\,422\,929\,868\,015\,284 = 2^2 \cdot 3^3 \cdot 7 \cdot 11 \cdot 19 \cdot 29 \cdot 31 \cdot 37 \cdot 53 \cdot 59 \cdot 61 \cdot 67 \cdot 71 \cdot 73$$

$$670\,475\,333\,050\,116\,177\,584 = 2^4 \cdot 7 \cdot 11 \cdot 13 \cdot 19 \cdot 29 \cdot 31 \cdot 37 \cdot 53 \cdot 59 \cdot 61 \cdot 67 \cdot 71 \cdot 73$$

$$1\,221\,222\,928\,055\,568\,752\,028 = 2^2 \cdot 3 \cdot 11 \cdot 13 \cdot 17 \cdot 19 \cdot 29 \cdot 31 \cdot 37 \cdot 53 \cdot 59 \cdot 61 \cdot 67 \cdot 71 \cdot 73$$

$$2\,105\,556\,772\,509\,601\,296\,600 = 2^3 \cdot 3 \cdot 5^2 \cdot 11 \cdot 13 \cdot 17 \cdot 19 \cdot 31 \cdot 37 \cdot 53 \cdot 59 \cdot 61 \cdot 67 \cdot 71 \cdot 73$$

$$3\,439\,076\,061\,765\,682\,117\,780 = 2^2 \cdot 5 \cdot 7^2 \cdot 11 \cdot 13 \cdot 17 \cdot 19 \cdot 31 \cdot 37 \cdot 53 \cdot 59 \cdot 61 \cdot 67 \cdot 71 \cdot 73$$

$$5\,325\,020\,998\,862\,991\,666\,240 = 2^6 \cdot 3 \cdot 5 \cdot 7^2 \cdot 11 \cdot 13 \cdot 17 \cdot 19 \cdot 37 \cdot 53 \cdot 59 \cdot 61 \cdot 67 \cdot 71 \cdot 73$$

$$7\,821\,124\,592\,080\,019\,009\,790 = 2 \cdot 3 \cdot 5 \cdot 7^2 \cdot 11 \cdot 13 \cdot 17 \cdot 19 \cdot 37 \cdot 47 \cdot 53 \cdot 59 \cdot 61 \cdot 67 \cdot 71 \cdot 73$$

$$10\,902\,173\,673\,808\,511\,346\,980 = 2^2 \cdot 5 \cdot 7^2 \cdot 13 \cdot 17 \cdot 19 \cdot 23 \cdot 37 \cdot 47 \cdot 53 \cdot 59 \cdot 61 \cdot 67 \cdot 71 \cdot 73$$

$$14\,429\,347\,509\,452\,441\,488\,650 = 2 \cdot 3^2 \cdot 5^2 \cdot 7^2 \cdot 13 \cdot 19 \cdot 23 \cdot 37 \cdot 47 \cdot 53 \cdot 59 \cdot 61 \cdot 67 \cdot 71 \cdot 73$$

$$18\,139\,751\,154\,740\,212\,157\,160 = 2^3 \cdot 3^2 \cdot 5 \cdot 7 \cdot 11 \cdot 13 \cdot 19 \cdot 23 \cdot 37 \cdot 47 \cdot 53 \cdot 59 \cdot 61 \cdot 67 \cdot 71 \cdot 73$$

$$21\,666\,924\,990\,384\,142\,298\,830 = 2 \cdot 5 \cdot 7 \cdot 11 \cdot 13 \cdot 19 \cdot 23 \cdot 37 \cdot 43 \cdot 47 \cdot 53 \cdot 59 \cdot 61 \cdot 67 \cdot 71 \cdot 73$$

$$24\,594\,887\,826\,922\,539\,906\,780 = 2^2 \cdot 3 \cdot 5 \cdot 7^2 \cdot 11 \cdot 13 \cdot 19 \cdot 23 \cdot 43 \cdot 47 \cdot 53 \cdot 59 \cdot 61 \cdot 67 \cdot 71 \cdot 73$$

$$26\,536\,589\,497\,469\,056\,215\,210 = 2 \cdot 3 \cdot 5 \cdot 7^2 \cdot 11 \cdot 13 \cdot 23 \cdot 41 \cdot 43 \cdot 47 \cdot 53 \cdot 59 \cdot 61 \cdot 67 \cdot 71 \cdot 73$$

$$27\,217\,014\,869\,199\,032\,015\,600 = 2^4 \cdot 5^2 \cdot 7^2 \cdot 11 \cdot 23 \cdot 41 \cdot 43 \cdot 47 \cdot 53 \cdot 59 \cdot 61 \cdot 67 \cdot 71 \cdot 73$$

Row 79

$$79 \text{ is Prime}$$

$$3\,081 = 3 \cdot 13 \cdot 79$$
$$79\,079 = 7 \cdot 11 \cdot 13 \cdot 79$$
$$1\,502\,501 = 7 \cdot 11 \cdot 13 \cdot 19 \cdot 79$$
$$22\,537\,515 = 3 \cdot 5 \cdot 7 \cdot 11 \cdot 13 \cdot 19 \cdot 79$$
$$277\,962\,685 = 5 \cdot 7 \cdot 11 \cdot 13 \cdot 19 \cdot 37 \cdot 79$$
$$2\,898\,753\,715 = 5 \cdot 11 \cdot 13 \cdot 19 \cdot 37 \cdot 73 \cdot 79$$
$$26\,088\,783\,435 = 3^2 \cdot 5 \cdot 11 \cdot 13 \cdot 19 \cdot 37 \cdot 73 \cdot 79$$
$$205\,811\,513\,765 = 5 \cdot 11 \cdot 13 \cdot 19 \cdot 37 \cdot 71 \cdot 73 \cdot 79$$
$$1\,440\,680\,596\,355 = 5 \cdot 7 \cdot 11 \cdot 13 \cdot 19 \cdot 37 \cdot 71 \cdot 73 \cdot 79$$
$$9\,036\,996\,468\,045 = 3 \cdot 5 \cdot 7 \cdot 13 \cdot 19 \cdot 23 \cdot 37 \cdot 71 \cdot 73 \cdot 79$$
$$51\,209\,646\,652\,255 = 5 \cdot 7 \cdot 13 \cdot 17 \cdot 19 \cdot 23 \cdot 37 \cdot 71 \cdot 73 \cdot 79$$
$$263\,926\,640\,438\,545 = 5 \cdot 7 \cdot 17 \cdot 19 \cdot 23 \cdot 37 \cdot 67 \cdot 71 \cdot 73 \cdot 79$$
$$1\,244\,225\,590\,638\,855 = 3 \cdot 5 \cdot 11 \cdot 17 \cdot 19 \cdot 23 \cdot 37 \cdot 67 \cdot 71 \cdot 73 \cdot 79$$
$$5\,391\,644\,226\,101\,705 = 5 \cdot 11 \cdot 13 \cdot 17 \cdot 19 \cdot 23 \cdot 37 \cdot 67 \cdot 71 \cdot 73 \cdot 79$$
$$21\,566\,576\,904\,406\,820 = 2^2 \cdot 5 \cdot 11 \cdot 13 \cdot 17 \cdot 19 \cdot 23 \cdot 37 \cdot 67 \cdot 71 \cdot 73 \cdot 79$$
$$79\,923\,196\,763\,389\,980 = 2^2 \cdot 3^2 \cdot 5 \cdot 7 \cdot 11 \cdot 13 \cdot 19 \cdot 23 \cdot 37 \cdot 67 \cdot 71 \cdot 73 \cdot 79$$
$$275\,291\,011\,073\,898\,820 = 2^2 \cdot 5 \cdot 7 \cdot 11 \cdot 13 \cdot 19 \cdot 23 \cdot 31 \cdot 37 \cdot 67 \cdot 71 \cdot 73 \cdot 79$$
$$883\,829\,035\,553\,043\,580 = 2^2 \cdot 5 \cdot 7 \cdot 11 \cdot 13 \cdot 23 \cdot 31 \cdot 37 \cdot 61 \cdot 67 \cdot 71 \cdot 73 \cdot 79$$
$$2\,651\,487\,106\,659\,130\,740 = 2^2 \cdot 3 \cdot 5 \cdot 7 \cdot 11 \cdot 13 \cdot 23 \cdot 31 \cdot 37 \cdot 61 \cdot 67 \cdot 71 \cdot 73 \cdot 79$$
$$7\,449\,416\,156\,804\,224\,460 = 2^2 \cdot 5 \cdot 11 \cdot 13 \cdot 23 \cdot 31 \cdot 37 \cdot 59 \cdot 61 \cdot 67 \cdot 71 \cdot 73 \cdot 79$$
$$19\,639\,369\,867\,938\,409\,940 = 2^2 \cdot 5 \cdot 13 \cdot 23 \cdot 29 \cdot 31 \cdot 37 \cdot 59 \cdot 61 \cdot 67 \cdot 71 \cdot 73 \cdot 79$$
$$48\,671\,481\,846\,629\,972\,460 = 2^2 \cdot 3 \cdot 5 \cdot 13 \cdot 19 \cdot 29 \cdot 31 \cdot 37 \cdot 59 \cdot 61 \cdot 67 \cdot 71 \cdot 73 \cdot 79$$
$$113\,566\,790\,975\,469\,935\,740 = 2^2 \cdot 5 \cdot 7 \cdot 13 \cdot 19 \cdot 29 \cdot 31 \cdot 37 \cdot 59 \cdot 61 \cdot 67 \cdot 71 \cdot 73 \cdot 79$$
$$249\,846\,940\,146\,033\,858\,628 = 2^2 \cdot 7 \cdot 11 \cdot 13 \cdot 19 \cdot 29 \cdot 31 \cdot 37 \cdot 59 \cdot 61 \cdot 67 \cdot 71 \cdot 73 \cdot 79$$
$$518\,912\,875\,687\,916\,475\,612 = 2^2 \cdot 3^3 \cdot 7 \cdot 11 \cdot 19 \cdot 29 \cdot 31 \cdot 37 \cdot 59 \cdot 61 \cdot 67 \cdot 71 \cdot 73 \cdot 79$$
$$1\,018\,606\,755\,979\,984\,192\,868 = 2^2 \cdot 7 \cdot 11 \cdot 19 \cdot 29 \cdot 31 \cdot 37 \cdot 53 \cdot 59 \cdot 61 \cdot 67 \cdot 71 \cdot 73 \cdot 79$$
$$1\,891\,698\,261\,105\,684\,929\,612 = 2^2 \cdot 11 \cdot 13 \cdot 19 \cdot 29 \cdot 31 \cdot 37 \cdot 53 \cdot 59 \cdot 61 \cdot 67 \cdot 71 \cdot 73 \cdot 79$$
$$3\,326\,779\,700\,565\,170\,048\,628 = 2^2 \cdot 3 \cdot 11 \cdot 13 \cdot 17 \cdot 19 \cdot 31 \cdot 37 \cdot 53 \cdot 59 \cdot 61 \cdot 67 \cdot 71 \cdot 73 \cdot 79$$
$$5\,544\,632\,834\,275\,283\,414\,380 = 2^2 \cdot 5 \cdot 11 \cdot 13 \cdot 17 \cdot 19 \cdot 31 \cdot 37 \cdot 53 \cdot 59 \cdot 61 \cdot 67 \cdot 71 \cdot 73 \cdot 79$$
$$8\,764\,097\,060\,628\,673\,784\,020 = 2^2 \cdot 5 \cdot 7^2 \cdot 11 \cdot 13 \cdot 17 \cdot 19 \cdot 37 \cdot 53 \cdot 59 \cdot 61 \cdot 67 \cdot 71 \cdot 73 \cdot 79$$
$$13\,146\,145\,590\,943\,010\,676\,030 = 2 \cdot 3 \cdot 5 \cdot 7^2 \cdot 11 \cdot 13 \cdot 17 \cdot 19 \cdot 37 \cdot 53 \cdot 59 \cdot 61 \cdot 67 \cdot 71 \cdot 73 \cdot 79$$
$$18\,723\,298\,265\,888\,530\,358\,770 = 2 \cdot 5 \cdot 7^2 \cdot 13 \cdot 17 \cdot 19 \cdot 37 \cdot 47 \cdot 53 \cdot 59 \cdot 61 \cdot 67 \cdot 71 \cdot 73 \cdot 79$$
$$25\,331\,521\,183\,260\,952\,835\,630 = 2 \cdot 5 \cdot 7^2 \cdot 13 \cdot 19 \cdot 23 \cdot 37 \cdot 47 \cdot 53 \cdot 59 \cdot 61 \cdot 67 \cdot 71 \cdot 73 \cdot 79$$
$$32\,569\,098\,664\,192\,653\,645\,810 = 2 \cdot 3^2 \cdot 5 \cdot 7 \cdot 13 \cdot 19 \cdot 23 \cdot 37 \cdot 47 \cdot 53 \cdot 59 \cdot 61 \cdot 67 \cdot 71 \cdot 73 \cdot 79$$
$$39\,806\,676\,145\,124\,354\,455\,990 = 2 \cdot 5 \cdot 7 \cdot 11 \cdot 13 \cdot 19 \cdot 23 \cdot 37 \cdot 47 \cdot 53 \cdot 59 \cdot 61 \cdot 67 \cdot 71 \cdot 73 \cdot 79$$
$$46\,261\,812\,817\,306\,682\,205\,610 = 2 \cdot 5 \cdot 7 \cdot 11 \cdot 13 \cdot 19 \cdot 23 \cdot 43 \cdot 47 \cdot 53 \cdot 59 \cdot 61 \cdot 67 \cdot 71 \cdot 73 \cdot 79$$
$$51\,131\,477\,324\,391\,596\,121\,990 = 2 \cdot 3 \cdot 5 \cdot 7^2 \cdot 11 \cdot 13 \cdot 23 \cdot 43 \cdot 47 \cdot 53 \cdot 59 \cdot 61 \cdot 67 \cdot 71 \cdot 73 \cdot 79$$
$$53\,753\,604\,366\,668\,088\,230\,810 = 2 \cdot 5 \cdot 7^2 \cdot 11 \cdot 23 \cdot 41 \cdot 43 \cdot 47 \cdot 53 \cdot 59 \cdot 61 \cdot 67 \cdot 71 \cdot 73 \cdot 79$$

Row 80

$$80 = 2^4 \cdot 5$$
$$3\,160 = 2^3 \cdot 5 \cdot 79$$
$$82\,160 = 2^4 \cdot 5 \cdot 13 \cdot 79$$
$$1\,581\,580 = 2^2 \cdot 5 \cdot 7 \cdot 11 \cdot 13 \cdot 79$$
$$24\,040\,016 = 2^4 \cdot 7 \cdot 11 \cdot 13 \cdot 19 \cdot 79$$
$$300\,500\,200 = 2^3 \cdot 5^2 \cdot 7 \cdot 11 \cdot 13 \cdot 19 \cdot 79$$
$$3\,176\,716\,400 = 2^4 \cdot 5^2 \cdot 11 \cdot 13 \cdot 19 \cdot 37 \cdot 79$$
$$28\,987\,537\,150 = 2 \cdot 5^2 \cdot 11 \cdot 13 \cdot 19 \cdot 37 \cdot 73 \cdot 79$$
$$231\,900\,297\,200 = 2^4 \cdot 5^2 \cdot 11 \cdot 13 \cdot 19 \cdot 37 \cdot 73 \cdot 79$$
$$1\,646\,492\,110\,120 = 2^3 \cdot 5 \cdot 11 \cdot 13 \cdot 19 \cdot 37 \cdot 71 \cdot 73 \cdot 79$$
$$10\,477\,677\,064\,400 = 2^4 \cdot 5^2 \cdot 7 \cdot 13 \cdot 19 \cdot 37 \cdot 71 \cdot 73 \cdot 79$$
$$60\,246\,643\,120\,300 = 2^2 \cdot 5^2 \cdot 7 \cdot 13 \cdot 19 \cdot 23 \cdot 37 \cdot 71 \cdot 73 \cdot 79$$
$$315\,136\,287\,090\,800 = 2^4 \cdot 5^2 \cdot 7 \cdot 17 \cdot 19 \cdot 23 \cdot 37 \cdot 71 \cdot 73 \cdot 79$$
$$1\,508\,152\,231\,077\,400 = 2^3 \cdot 5^2 \cdot 17 \cdot 19 \cdot 23 \cdot 37 \cdot 67 \cdot 71 \cdot 73 \cdot 79$$
$$6\,635\,869\,816\,740\,560 = 2^4 \cdot 5 \cdot 11 \cdot 17 \cdot 19 \cdot 23 \cdot 37 \cdot 67 \cdot 71 \cdot 73 \cdot 79$$
$$26\,958\,221\,130\,508\,525 = 5^2 \cdot 11 \cdot 13 \cdot 17 \cdot 19 \cdot 23 \cdot 37 \cdot 67 \cdot 71 \cdot 73 \cdot 79$$
$$101\,489\,773\,667\,796\,800 = 2^6 \cdot 5^2 \cdot 11 \cdot 13 \cdot 19 \cdot 23 \cdot 37 \cdot 67 \cdot 71 \cdot 73 \cdot 79$$
$$355\,214\,207\,837\,288\,800 = 2^5 \cdot 5^2 \cdot 7 \cdot 11 \cdot 13 \cdot 19 \cdot 23 \cdot 37 \cdot 67 \cdot 71 \cdot 73 \cdot 79$$
$$1\,159\,120\,046\,626\,942\,400 = 2^6 \cdot 5^2 \cdot 7 \cdot 11 \cdot 13 \cdot 23 \cdot 31 \cdot 37 \cdot 67 \cdot 71 \cdot 73 \cdot 79$$
$$3\,535\,316\,142\,212\,174\,320 = 2^4 \cdot 5 \cdot 7 \cdot 11 \cdot 13 \cdot 23 \cdot 31 \cdot 37 \cdot 61 \cdot 67 \cdot 71 \cdot 73 \cdot 79$$
$$10\,100\,903\,263\,463\,355\,200 = 2^6 \cdot 5^2 \cdot 11 \cdot 13 \cdot 23 \cdot 31 \cdot 37 \cdot 61 \cdot 67 \cdot 71 \cdot 73 \cdot 79$$
$$27\,088\,786\,024\,742\,634\,400 = 2^5 \cdot 5^2 \cdot 13 \cdot 23 \cdot 31 \cdot 37 \cdot 59 \cdot 61 \cdot 67 \cdot 71 \cdot 73 \cdot 79$$
$$68\,310\,851\,714\,568\,382\,400 = 2^6 \cdot 5^2 \cdot 13 \cdot 29 \cdot 31 \cdot 37 \cdot 59 \cdot 61 \cdot 67 \cdot 71 \cdot 73 \cdot 79$$
$$162\,238\,272\,822\,099\,908\,200 = 2^3 \cdot 5^2 \cdot 13 \cdot 19 \cdot 29 \cdot 31 \cdot 37 \cdot 59 \cdot 61 \cdot 67 \cdot 71 \cdot 73 \cdot 79$$
$$363\,413\,731\,121\,503\,794\,368 = 2^6 \cdot 7 \cdot 13 \cdot 19 \cdot 29 \cdot 31 \cdot 37 \cdot 59 \cdot 61 \cdot 67 \cdot 71 \cdot 73 \cdot 79$$
$$768\,759\,815\,833\,950\,334\,240 = 2^5 \cdot 5 \cdot 7 \cdot 11 \cdot 19 \cdot 29 \cdot 31 \cdot 37 \cdot 59 \cdot 61 \cdot 67 \cdot 71 \cdot 73 \cdot 79$$
$$1\,537\,519\,631\,667\,900\,668\,480 = 2^6 \cdot 5 \cdot 7 \cdot 11 \cdot 19 \cdot 29 \cdot 31 \cdot 37 \cdot 59 \cdot 61 \cdot 67 \cdot 71 \cdot 73 \cdot 79$$
$$2\,910\,305\,017\,085\,669\,122\,480 = 2^4 \cdot 5 \cdot 11 \cdot 19 \cdot 29 \cdot 31 \cdot 37 \cdot 53 \cdot 59 \cdot 61 \cdot 67 \cdot 71 \cdot 73 \cdot 79$$
$$5\,218\,477\,961\,670\,854\,978\,240 = 2^6 \cdot 5 \cdot 11 \cdot 13 \cdot 19 \cdot 31 \cdot 37 \cdot 53 \cdot 59 \cdot 61 \cdot 67 \cdot 71 \cdot 73 \cdot 79$$
$$8\,871\,412\,534\,840\,453\,463\,008 = 2^5 \cdot 11 \cdot 13 \cdot 17 \cdot 19 \cdot 31 \cdot 37 \cdot 53 \cdot 59 \cdot 61 \cdot 67 \cdot 71 \cdot 73 \cdot 79$$
$$14\,308\,729\,894\,903\,957\,198\,400 = 2^6 \cdot 5^2 \cdot 11 \cdot 13 \cdot 17 \cdot 19 \cdot 37 \cdot 53 \cdot 59 \cdot 61 \cdot 67 \cdot 71 \cdot 73 \cdot 79$$
$$21\,910\,242\,651\,571\,684\,460\,050 = 2 \cdot 5^2 \cdot 7^2 \cdot 11 \cdot 13 \cdot 17 \cdot 19 \cdot 37 \cdot 53 \cdot 59 \cdot 61 \cdot 67 \cdot 71 \cdot 73 \cdot 79$$
$$31\,869\,443\,856\,831\,541\,032\,800 = 2^5 \cdot 5^2 \cdot 7^2 \cdot 13 \cdot 17 \cdot 19 \cdot 37 \cdot 53 \cdot 59 \cdot 61 \cdot 67 \cdot 71 \cdot 73 \cdot 79$$
$$44\,054\,819\,449\,149\,483\,192\,400 = 2^4 \cdot 5^2 \cdot 7^2 \cdot 13 \cdot 19 \cdot 37 \cdot 47 \cdot 53 \cdot 59 \cdot 61 \cdot 67 \cdot 71 \cdot 73 \cdot 79$$
$$57\,900\,619\,847\,453\,606\,481\,440 = 2^5 \cdot 5 \cdot 7 \cdot 13 \cdot 19 \cdot 23 \cdot 37 \cdot 47 \cdot 53 \cdot 59 \cdot 61 \cdot 67 \cdot 71 \cdot 73 \cdot 79$$
$$72\,375\,774\,809\,317\,008\,101\,800 = 2^3 \cdot 5^2 \cdot 7 \cdot 13 \cdot 19 \cdot 23 \cdot 37 \cdot 47 \cdot 53 \cdot 59 \cdot 61 \cdot 67 \cdot 71 \cdot 73 \cdot 79$$
$$86\,068\,488\,962\,431\,036\,661\,600 = 2^5 \cdot 5^2 \cdot 7 \cdot 11 \cdot 13 \cdot 19 \cdot 23 \cdot 47 \cdot 53 \cdot 59 \cdot 61 \cdot 67 \cdot 71 \cdot 73 \cdot 79$$
$$97\,393\,290\,141\,698\,278\,327\,600 = 2^4 \cdot 5^2 \cdot 7 \cdot 11 \cdot 13 \cdot 23 \cdot 43 \cdot 47 \cdot 53 \cdot 59 \cdot 61 \cdot 67 \cdot 71 \cdot 73 \cdot 79$$
$$104\,885\,081\,691\,059\,684\,352\,800 = 2^5 \cdot 5^2 \cdot 7^2 \cdot 11 \cdot 23 \cdot 43 \cdot 47 \cdot 53 \cdot 59 \cdot 61 \cdot 67 \cdot 71 \cdot 73 \cdot 79$$
$$107\,507\,208\,733\,336\,176\,461\,620 = 2^2 \cdot 5 \cdot 7^2 \cdot 11 \cdot 23 \cdot 41 \cdot 43 \cdot 47 \cdot 53 \cdot 59 \cdot 61 \cdot 67 \cdot 71 \cdot 73 \cdot 79$$

Pascal's Triangle — Prime Factorization — To Center Number (omitting 1's)

Row 81

$$81 = 3^4$$
$$3\,240 = 2^3 \cdot 3^4 \cdot 5$$
$$85\,320 = 2^3 \cdot 3^3 \cdot 5 \cdot 79$$
$$1\,663\,740 = 2^2 \cdot 3^4 \cdot 5 \cdot 13 \cdot 79$$
$$25\,621\,596 = 2^2 \cdot 3^4 \cdot 7 \cdot 11 \cdot 13 \cdot 79$$
$$324\,540\,216 = 2^3 \cdot 3^3 \cdot 7 \cdot 11 \cdot 13 \cdot 19 \cdot 79$$
$$3\,477\,216\,600 = 2^3 \cdot 3^4 \cdot 5^2 \cdot 11 \cdot 13 \cdot 19 \cdot 79$$
$$32\,164\,253\,550 = 2 \cdot 3^4 \cdot 5^2 \cdot 11 \cdot 13 \cdot 19 \cdot 37 \cdot 79$$
$$260\,887\,834\,350 = 2 \cdot 3^2 \cdot 5^2 \cdot 11 \cdot 13 \cdot 19 \cdot 37 \cdot 73 \cdot 79$$
$$1\,878\,392\,407\,320 = 2^3 \cdot 3^4 \cdot 5 \cdot 11 \cdot 13 \cdot 19 \cdot 37 \cdot 73 \cdot 79$$
$$12\,124\,169\,174\,520 = 2^3 \cdot 3^4 \cdot 5 \cdot 13 \cdot 19 \cdot 37 \cdot 71 \cdot 73 \cdot 79$$
$$70\,724\,320\,184\,700 = 2^2 \cdot 3^3 \cdot 5^2 \cdot 7 \cdot 13 \cdot 19 \cdot 37 \cdot 71 \cdot 73 \cdot 79$$
$$375\,382\,930\,211\,100 = 2^2 \cdot 3^4 \cdot 5^2 \cdot 7 \cdot 19 \cdot 23 \cdot 37 \cdot 71 \cdot 73 \cdot 79$$
$$1\,823\,288\,518\,168\,200 = 2^3 \cdot 3^4 \cdot 5^2 \cdot 17 \cdot 19 \cdot 23 \cdot 37 \cdot 71 \cdot 73 \cdot 79$$
$$8\,144\,022\,047\,817\,960 = 2^3 \cdot 3^3 \cdot 5 \cdot 17 \cdot 19 \cdot 23 \cdot 37 \cdot 67 \cdot 71 \cdot 73 \cdot 79$$
$$33\,594\,090\,947\,249\,085 = 3^4 \cdot 5 \cdot 11 \cdot 17 \cdot 19 \cdot 23 \cdot 37 \cdot 67 \cdot 71 \cdot 73 \cdot 79$$
$$128\,447\,994\,798\,305\,325 = 3^4 \cdot 5^2 \cdot 11 \cdot 13 \cdot 19 \cdot 23 \cdot 37 \cdot 67 \cdot 71 \cdot 73 \cdot 79$$
$$456\,703\,981\,505\,085\,600 = 2^5 \cdot 3^2 \cdot 5^2 \cdot 11 \cdot 13 \cdot 19 \cdot 23 \cdot 37 \cdot 67 \cdot 71 \cdot 73 \cdot 79$$
$$1\,514\,334\,254\,464\,231\,200 = 2^5 \cdot 3^4 \cdot 5^2 \cdot 7 \cdot 11 \cdot 13 \cdot 23 \cdot 37 \cdot 67 \cdot 71 \cdot 73 \cdot 79$$
$$4\,694\,436\,188\,839\,116\,720 = 2^4 \cdot 3^4 \cdot 5 \cdot 7 \cdot 11 \cdot 13 \cdot 23 \cdot 31 \cdot 37 \cdot 67 \cdot 71 \cdot 73 \cdot 79$$
$$13\,636\,219\,405\,675\,529\,520 = 2^4 \cdot 3^3 \cdot 5 \cdot 11 \cdot 13 \cdot 23 \cdot 31 \cdot 37 \cdot 61 \cdot 67 \cdot 71 \cdot 73 \cdot 79$$
$$37\,189\,689\,288\,205\,989\,600 = 2^5 \cdot 3^4 \cdot 5^2 \cdot 13 \cdot 23 \cdot 31 \cdot 37 \cdot 61 \cdot 67 \cdot 71 \cdot 73 \cdot 79$$
$$95\,399\,637\,739\,311\,016\,800 = 2^5 \cdot 3^4 \cdot 5^2 \cdot 13 \cdot 31 \cdot 37 \cdot 59 \cdot 61 \cdot 67 \cdot 71 \cdot 73 \cdot 79$$
$$230\,549\,124\,536\,668\,290\,600 = 2^3 \cdot 3^3 \cdot 5^2 \cdot 13 \cdot 29 \cdot 31 \cdot 37 \cdot 59 \cdot 61 \cdot 67 \cdot 71 \cdot 73 \cdot 79$$
$$525\,652\,003\,943\,603\,702\,568 = 2^3 \cdot 3^4 \cdot 13 \cdot 19 \cdot 29 \cdot 31 \cdot 37 \cdot 59 \cdot 61 \cdot 67 \cdot 71 \cdot 73 \cdot 79$$
$$1\,132\,173\,546\,955\,454\,128\,608 = 2^5 \cdot 3^4 \cdot 7 \cdot 19 \cdot 29 \cdot 31 \cdot 37 \cdot 59 \cdot 61 \cdot 67 \cdot 71 \cdot 73 \cdot 79$$
$$2\,306\,279\,447\,501\,851\,002\,720 = 2^5 \cdot 3 \cdot 5 \cdot 7 \cdot 11 \cdot 19 \cdot 29 \cdot 31 \cdot 37 \cdot 59 \cdot 61 \cdot 67 \cdot 71 \cdot 73 \cdot 79$$
$$4\,447\,824\,648\,753\,569\,790\,960 = 2^4 \cdot 3^4 \cdot 5 \cdot 11 \cdot 19 \cdot 29 \cdot 31 \cdot 37 \cdot 59 \cdot 61 \cdot 67 \cdot 71 \cdot 73 \cdot 79$$
$$8\,128\,782\,978\,756\,524\,100\,720 = 2^4 \cdot 3^4 \cdot 5 \cdot 11 \cdot 19 \cdot 31 \cdot 37 \cdot 53 \cdot 59 \cdot 61 \cdot 67 \cdot 71 \cdot 73 \cdot 79$$
$$14\,089\,890\,496\,511\,308\,441\,248 = 2^5 \cdot 3^3 \cdot 11 \cdot 13 \cdot 19 \cdot 31 \cdot 37 \cdot 53 \cdot 59 \cdot 61 \cdot 67 \cdot 71 \cdot 73 \cdot 79$$
$$23\,180\,142\,429\,744\,410\,661\,408 = 2^5 \cdot 3^4 \cdot 11 \cdot 13 \cdot 17 \cdot 19 \cdot 37 \cdot 53 \cdot 59 \cdot 61 \cdot 67 \cdot 71 \cdot 73 \cdot 79$$
$$36\,218\,972\,546\,475\,641\,658\,450 = 2 \cdot 3^4 \cdot 5^2 \cdot 11 \cdot 13 \cdot 17 \cdot 19 \cdot 37 \cdot 53 \cdot 59 \cdot 61 \cdot 67 \cdot 71 \cdot 73 \cdot 79$$
$$53\,779\,686\,508\,403\,225\,492\,850 = 2 \cdot 3^3 \cdot 5^2 \cdot 7^2 \cdot 13 \cdot 17 \cdot 19 \cdot 37 \cdot 53 \cdot 59 \cdot 61 \cdot 67 \cdot 71 \cdot 73 \cdot 79$$
$$75\,924\,263\,305\,981\,024\,225\,200 = 2^4 \cdot 3^4 \cdot 5^2 \cdot 7^2 \cdot 13 \cdot 19 \cdot 37 \cdot 53 \cdot 59 \cdot 61 \cdot 67 \cdot 71 \cdot 73 \cdot 79$$
$$101\,955\,439\,296\,603\,089\,673\,840 = 2^4 \cdot 3^4 \cdot 5 \cdot 7 \cdot 13 \cdot 19 \cdot 37 \cdot 47 \cdot 53 \cdot 59 \cdot 61 \cdot 67 \cdot 71 \cdot 73 \cdot 79$$
$$130\,276\,394\,656\,770\,614\,583\,240 = 2^3 \cdot 3^2 \cdot 5 \cdot 7 \cdot 13 \cdot 19 \cdot 23 \cdot 37 \cdot 47 \cdot 53 \cdot 59 \cdot 61 \cdot 67 \cdot 71 \cdot 73 \cdot 79$$
$$158\,444\,263\,771\,748\,044\,763\,400 = 2^3 \cdot 3^4 \cdot 5^2 \cdot 7 \cdot 13 \cdot 19 \cdot 23 \cdot 47 \cdot 53 \cdot 59 \cdot 61 \cdot 67 \cdot 71 \cdot 73 \cdot 79$$
$$183\,461\,779\,104\,129\,314\,989\,200 = 2^4 \cdot 3^4 \cdot 5^2 \cdot 7 \cdot 11 \cdot 13 \cdot 23 \cdot 47 \cdot 53 \cdot 59 \cdot 61 \cdot 67 \cdot 71 \cdot 73 \cdot 79$$
$$202\,278\,371\,832\,757\,962\,680\,400 = 2^4 \cdot 3^3 \cdot 5^2 \cdot 7 \cdot 11 \cdot 23 \cdot 43 \cdot 47 \cdot 53 \cdot 59 \cdot 61 \cdot 67 \cdot 71 \cdot 73 \cdot 79$$
$$212\,392\,290\,424\,395\,860\,814\,420 = 2^2 \cdot 3^4 \cdot 5 \cdot 7^2 \cdot 11 \cdot 23 \cdot 43 \cdot 47 \cdot 53 \cdot 59 \cdot 61 \cdot 67 \cdot 71 \cdot 73 \cdot 79$$

Row 82

$$82 = 2 \cdot 41$$

$$3\,321 = 3^4 \cdot 41$$

$$88\,560 = 2^4 \cdot 3^3 \cdot 5 \cdot 41$$

$$1\,749\,060 = 2^2 \cdot 3^3 \cdot 5 \cdot 41 \cdot 79$$

$$27\,285\,336 = 2^3 \cdot 3^4 \cdot 13 \cdot 41 \cdot 79$$

$$350\,161\,812 = 2^2 \cdot 3^3 \cdot 7 \cdot 11 \cdot 13 \cdot 41 \cdot 79$$

$$3\,801\,756\,816 = 2^4 \cdot 3^3 \cdot 11 \cdot 13 \cdot 19 \cdot 41 \cdot 79$$

$$35\,641\,470\,150 = 2 \cdot 3^4 \cdot 5^2 \cdot 11 \cdot 13 \cdot 19 \cdot 41 \cdot 79$$

$$293\,052\,087\,900 = 2^2 \cdot 3^2 \cdot 5^2 \cdot 11 \cdot 13 \cdot 19 \cdot 37 \cdot 41 \cdot 79$$

$$2\,139\,280\,241\,670 = 2 \cdot 3^2 \cdot 5 \cdot 11 \cdot 13 \cdot 19 \cdot 37 \cdot 41 \cdot 73 \cdot 79$$

$$14\,002\,561\,581\,840 = 2^4 \cdot 3^4 \cdot 5 \cdot 13 \cdot 19 \cdot 37 \cdot 41 \cdot 73 \cdot 79$$

$$82\,848\,489\,359\,220 = 2^2 \cdot 3^3 \cdot 5 \cdot 13 \cdot 19 \cdot 37 \cdot 41 \cdot 71 \cdot 73 \cdot 79$$

$$446\,107\,250\,395\,800 = 2^3 \cdot 3^3 \cdot 5^2 \cdot 7 \cdot 19 \cdot 37 \cdot 41 \cdot 71 \cdot 73 \cdot 79$$

$$2\,198\,671\,448\,379\,300 = 2^2 \cdot 3^4 \cdot 5^2 \cdot 19 \cdot 23 \cdot 37 \cdot 41 \cdot 71 \cdot 73 \cdot 79$$

$$9\,967\,310\,565\,986\,160 = 2^4 \cdot 3^3 \cdot 5 \cdot 17 \cdot 19 \cdot 23 \cdot 37 \cdot 41 \cdot 71 \cdot 73 \cdot 79$$

$$41\,738\,112\,995\,067\,045 = 3^3 \cdot 5 \cdot 17 \cdot 19 \cdot 23 \cdot 37 \cdot 41 \cdot 67 \cdot 71 \cdot 73 \cdot 79$$

$$162\,042\,085\,745\,554\,410 = 2 \cdot 3^4 \cdot 5 \cdot 11 \cdot 19 \cdot 23 \cdot 37 \cdot 41 \cdot 67 \cdot 71 \cdot 73 \cdot 79$$

$$585\,151\,976\,303\,390\,925 = 3^2 \cdot 5^2 \cdot 11 \cdot 13 \cdot 19 \cdot 23 \cdot 37 \cdot 41 \cdot 67 \cdot 71 \cdot 73 \cdot 79$$

$$1\,971\,038\,235\,969\,316\,800 = 2^6 \cdot 3^2 \cdot 5^2 \cdot 11 \cdot 13 \cdot 23 \cdot 37 \cdot 41 \cdot 67 \cdot 71 \cdot 73 \cdot 79$$

$$6\,208\,770\,443\,303\,347\,920 = 2^4 \cdot 3^4 \cdot 5 \cdot 7 \cdot 11 \cdot 13 \cdot 23 \cdot 37 \cdot 41 \cdot 67 \cdot 71 \cdot 73 \cdot 79$$

$$18\,330\,655\,594\,514\,646\,240 = 2^5 \cdot 3^3 \cdot 5 \cdot 11 \cdot 13 \cdot 23 \cdot 31 \cdot 37 \cdot 41 \cdot 67 \cdot 71 \cdot 73 \cdot 79$$

$$50\,825\,908\,693\,881\,519\,120 = 2^4 \cdot 3^3 \cdot 5 \cdot 13 \cdot 23 \cdot 31 \cdot 37 \cdot 41 \cdot 61 \cdot 67 \cdot 71 \cdot 73 \cdot 79$$

$$132\,589\,327\,027\,517\,006\,400 = 2^6 \cdot 3^4 \cdot 5^2 \cdot 13 \cdot 31 \cdot 37 \cdot 41 \cdot 61 \cdot 67 \cdot 71 \cdot 73 \cdot 79$$

$$325\,948\,762\,275\,979\,307\,400 = 2^3 \cdot 3^3 \cdot 5^2 \cdot 13 \cdot 31 \cdot 37 \cdot 41 \cdot 59 \cdot 61 \cdot 67 \cdot 71 \cdot 73 \cdot 79$$

$$756\,201\,128\,480\,271\,993\,168 = 2^4 \cdot 3^3 \cdot 13 \cdot 29 \cdot 31 \cdot 37 \cdot 41 \cdot 59 \cdot 61 \cdot 67 \cdot 71 \cdot 73 \cdot 79$$

$$1\,657\,825\,550\,899\,057\,831\,176 = 2^3 \cdot 3^4 \cdot 19 \cdot 29 \cdot 31 \cdot 37 \cdot 41 \cdot 59 \cdot 61 \cdot 67 \cdot 71 \cdot 73 \cdot 79$$

$$3\,438\,452\,994\,457\,305\,131\,328 = 2^6 \cdot 3 \cdot 7 \cdot 19 \cdot 29 \cdot 31 \cdot 37 \cdot 41 \cdot 59 \cdot 61 \cdot 67 \cdot 71 \cdot 73 \cdot 79$$

$$6\,754\,104\,096\,255\,420\,793\,680 = 2^4 \cdot 3 \cdot 5 \cdot 11 \cdot 19 \cdot 29 \cdot 31 \cdot 37 \cdot 41 \cdot 59 \cdot 61 \cdot 67 \cdot 71 \cdot 73 \cdot 79$$

$$12\,576\,607\,627\,510\,093\,891\,680 = 2^5 \cdot 3^4 \cdot 5 \cdot 11 \cdot 19 \cdot 31 \cdot 37 \cdot 41 \cdot 59 \cdot 61 \cdot 67 \cdot 71 \cdot 73 \cdot 79$$

$$22\,218\,673\,475\,267\,832\,541\,968 = 2^4 \cdot 3^3 \cdot 11 \cdot 19 \cdot 31 \cdot 37 \cdot 41 \cdot 53 \cdot 59 \cdot 61 \cdot 67 \cdot 71 \cdot 73 \cdot 79$$

$$37\,270\,032\,926\,255\,719\,102\,656 = 2^6 \cdot 3^3 \cdot 11 \cdot 13 \cdot 19 \cdot 37 \cdot 41 \cdot 53 \cdot 59 \cdot 61 \cdot 67 \cdot 71 \cdot 73 \cdot 79$$

$$59\,399\,114\,976\,220\,052\,319\,858 = 2 \cdot 3^4 \cdot 11 \cdot 13 \cdot 17 \cdot 19 \cdot 37 \cdot 41 \cdot 53 \cdot 59 \cdot 61 \cdot 67 \cdot 71 \cdot 73 \cdot 79$$

$$89\,998\,659\,054\,878\,867\,151\,300 = 2^2 \cdot 3^3 \cdot 5^2 \cdot 13 \cdot 17 \cdot 19 \cdot 37 \cdot 41 \cdot 53 \cdot 59 \cdot 61 \cdot 67 \cdot 71 \cdot 73 \cdot 79$$

$$129\,703\,949\,814\,384\,249\,718\,050 = 2 \cdot 3^3 \cdot 5^2 \cdot 7^2 \cdot 13 \cdot 19 \cdot 37 \cdot 41 \cdot 53 \cdot 59 \cdot 61 \cdot 67 \cdot 71 \cdot 73 \cdot 79$$

$$177\,879\,702\,602\,584\,113\,899\,040 = 2^5 \cdot 3^4 \cdot 5 \cdot 7 \cdot 13 \cdot 19 \cdot 37 \cdot 41 \cdot 53 \cdot 59 \cdot 61 \cdot 67 \cdot 71 \cdot 73 \cdot 79$$

$$232\,231\,833\,953\,373\,704\,257\,080 = 2^3 \cdot 3^2 \cdot 5 \cdot 7 \cdot 13 \cdot 19 \cdot 37 \cdot 41 \cdot 47 \cdot 53 \cdot 59 \cdot 61 \cdot 67 \cdot 71 \cdot 73 \cdot 79$$

$$288\,720\,658\,428\,518\,659\,346\,640 = 2^4 \cdot 3^2 \cdot 5 \cdot 7 \cdot 13 \cdot 19 \cdot 23 \cdot 41 \cdot 47 \cdot 53 \cdot 59 \cdot 61 \cdot 67 \cdot 71 \cdot 73 \cdot 79$$

$$341\,906\,042\,875\,877\,359\,752\,600 = 2^3 \cdot 3^4 \cdot 5^2 \cdot 7 \cdot 13 \cdot 23 \cdot 41 \cdot 47 \cdot 53 \cdot 59 \cdot 61 \cdot 67 \cdot 71 \cdot 73 \cdot 79$$

$$385\,740\,150\,936\,887\,277\,669\,600 = 2^5 \cdot 3^3 \cdot 5^2 \cdot 7 \cdot 11 \cdot 23 \cdot 41 \cdot 47 \cdot 53 \cdot 59 \cdot 61 \cdot 67 \cdot 71 \cdot 73 \cdot 79$$

$$414\,670\,662\,257\,153\,823\,494\,820 = 2^2 \cdot 3^3 \cdot 5 \cdot 7 \cdot 11 \cdot 23 \cdot 41 \cdot 43 \cdot 47 \cdot 53 \cdot 59 \cdot 61 \cdot 67 \cdot 71 \cdot 73 \cdot 79$$

$$424\,784\,580\,848\,791\,721\,628\,840 = 2^3 \cdot 3^4 \cdot 5 \cdot 7^2 \cdot 11 \cdot 23 \cdot 43 \cdot 47 \cdot 53 \cdot 59 \cdot 61 \cdot 67 \cdot 71 \cdot 73 \cdot 79$$

Row 83

$$
\begin{aligned}
83 &\ \text{is Prime} \\
3\,403 &= 41 \cdot 83 \\
91\,881 &= 3^3 \cdot 41 \cdot 83 \\
1\,837\,620 &= 2^2 \cdot 3^3 \cdot 5 \cdot 41 \cdot 83 \\
29\,034\,396 &= 2^2 \cdot 3^3 \cdot 41 \cdot 79 \cdot 83 \\
377\,447\,148 &= 2^2 \cdot 3^3 \cdot 13 \cdot 41 \cdot 79 \cdot 83 \\
4\,151\,918\,628 &= 2^2 \cdot 3^3 \cdot 11 \cdot 13 \cdot 41 \cdot 79 \cdot 83 \\
39\,443\,226\,966 &= 2 \cdot 3^3 \cdot 11 \cdot 13 \cdot 19 \cdot 41 \cdot 79 \cdot 83 \\
328\,693\,558\,050 &= 2 \cdot 3^2 \cdot 5^2 \cdot 11 \cdot 13 \cdot 19 \cdot 41 \cdot 79 \cdot 83 \\
2\,432\,332\,329\,570 &= 2 \cdot 3^2 \cdot 5 \cdot 11 \cdot 13 \cdot 19 \cdot 37 \cdot 41 \cdot 79 \cdot 83 \\
16\,141\,841\,823\,510 &= 2 \cdot 3^2 \cdot 5 \cdot 13 \cdot 19 \cdot 37 \cdot 41 \cdot 73 \cdot 79 \cdot 83 \\
96\,851\,050\,941\,060 &= 2^2 \cdot 3^3 \cdot 5 \cdot 13 \cdot 19 \cdot 37 \cdot 41 \cdot 73 \cdot 79 \cdot 83 \\
528\,955\,739\,755\,020 &= 2^2 \cdot 3^3 \cdot 5 \cdot 19 \cdot 37 \cdot 41 \cdot 71 \cdot 73 \cdot 79 \cdot 83 \\
2\,644\,778\,698\,775\,100 &= 2^2 \cdot 3^3 \cdot 5^2 \cdot 19 \cdot 37 \cdot 41 \cdot 71 \cdot 73 \cdot 79 \cdot 83 \\
12\,165\,982\,014\,365\,460 &= 2^2 \cdot 3^3 \cdot 5 \cdot 19 \cdot 23 \cdot 37 \cdot 41 \cdot 71 \cdot 73 \cdot 79 \cdot 83 \\
51\,705\,423\,561\,053\,205 &= 3^3 \cdot 5 \cdot 17 \cdot 19 \cdot 23 \cdot 37 \cdot 41 \cdot 71 \cdot 73 \cdot 79 \cdot 83 \\
203\,780\,198\,740\,621\,455 &= 3^3 \cdot 5 \cdot 19 \cdot 23 \cdot 37 \cdot 41 \cdot 67 \cdot 71 \cdot 73 \cdot 79 \cdot 83 \\
747\,194\,062\,048\,945\,335 &= 3^2 \cdot 5 \cdot 11 \cdot 19 \cdot 23 \cdot 37 \cdot 41 \cdot 67 \cdot 71 \cdot 73 \cdot 79 \cdot 83 \\
2\,556\,190\,212\,272\,707\,725 &= 3^2 \cdot 5^2 \cdot 11 \cdot 13 \cdot 23 \cdot 37 \cdot 41 \cdot 67 \cdot 71 \cdot 73 \cdot 79 \cdot 83 \\
8\,179\,808\,679\,272\,664\,720 &= 2^4 \cdot 3^2 \cdot 5 \cdot 11 \cdot 13 \cdot 23 \cdot 37 \cdot 41 \cdot 67 \cdot 71 \cdot 73 \cdot 79 \cdot 83 \\
24\,539\,426\,037\,817\,994\,160 &= 2^4 \cdot 3^3 \cdot 5 \cdot 11 \cdot 13 \cdot 23 \cdot 37 \cdot 41 \cdot 67 \cdot 71 \cdot 73 \cdot 79 \cdot 83 \\
69\,156\,564\,288\,396\,165\,360 &= 2^4 \cdot 3^3 \cdot 5 \cdot 13 \cdot 23 \cdot 31 \cdot 37 \cdot 41 \cdot 67 \cdot 71 \cdot 73 \cdot 79 \cdot 83 \\
183\,415\,235\,721\,398\,525\,520 &= 2^4 \cdot 3^3 \cdot 5 \cdot 13 \cdot 31 \cdot 37 \cdot 41 \cdot 61 \cdot 67 \cdot 71 \cdot 73 \cdot 79 \cdot 83 \\
458\,538\,089\,303\,496\,313\,800 &= 2^3 \cdot 3^3 \cdot 5^2 \cdot 13 \cdot 31 \cdot 37 \cdot 41 \cdot 61 \cdot 67 \cdot 71 \cdot 73 \cdot 79 \cdot 83 \\
1\,082\,149\,890\,756\,251\,300\,568 &= 2^3 \cdot 3^3 \cdot 13 \cdot 31 \cdot 37 \cdot 41 \cdot 59 \cdot 61 \cdot 67 \cdot 71 \cdot 73 \cdot 79 \cdot 83 \\
2\,414\,026\,679\,379\,329\,824\,344 &= 2^3 \cdot 3^3 \cdot 29 \cdot 31 \cdot 37 \cdot 41 \cdot 59 \cdot 61 \cdot 67 \cdot 71 \cdot 73 \cdot 79 \cdot 83 \\
5\,096\,278\,545\,356\,362\,962\,504 &= 2^3 \cdot 3 \cdot 19 \cdot 29 \cdot 31 \cdot 37 \cdot 41 \cdot 59 \cdot 61 \cdot 67 \cdot 71 \cdot 73 \cdot 79 \cdot 83 \\
10\,192\,557\,090\,712\,725\,925\,008 &= 2^4 \cdot 3 \cdot 19 \cdot 29 \cdot 31 \cdot 37 \cdot 41 \cdot 59 \cdot 61 \cdot 67 \cdot 71 \cdot 73 \cdot 79 \cdot 83 \\
19\,330\,711\,723\,765\,514\,685\,360 &= 2^4 \cdot 3 \cdot 5 \cdot 11 \cdot 19 \cdot 31 \cdot 37 \cdot 41 \cdot 59 \cdot 61 \cdot 67 \cdot 71 \cdot 73 \cdot 79 \cdot 83 \\
34\,795\,281\,102\,777\,926\,433\,648 &= 2^4 \cdot 3^3 \cdot 11 \cdot 19 \cdot 31 \cdot 37 \cdot 41 \cdot 59 \cdot 61 \cdot 67 \cdot 71 \cdot 73 \cdot 79 \cdot 83 \\
59\,488\,706\,401\,523\,551\,644\,624 &= 2^4 \cdot 3^3 \cdot 11 \cdot 19 \cdot 37 \cdot 41 \cdot 53 \cdot 59 \cdot 61 \cdot 67 \cdot 71 \cdot 73 \cdot 79 \cdot 83 \\
96\,669\,147\,902\,475\,771\,422\,514 &= 2 \cdot 3^3 \cdot 11 \cdot 13 \cdot 19 \cdot 37 \cdot 41 \cdot 53 \cdot 59 \cdot 61 \cdot 67 \cdot 71 \cdot 73 \cdot 79 \cdot 83 \\
149\,397\,774\,031\,098\,919\,471\,158 &= 2 \cdot 3^3 \cdot 13 \cdot 17 \cdot 19 \cdot 37 \cdot 41 \cdot 53 \cdot 59 \cdot 61 \cdot 67 \cdot 71 \cdot 73 \cdot 79 \cdot 83 \\
219\,702\,608\,869\,263\,116\,869\,350 &= 2 \cdot 3^3 \cdot 5^2 \cdot 13 \cdot 19 \cdot 37 \cdot 41 \cdot 53 \cdot 59 \cdot 61 \cdot 67 \cdot 71 \cdot 73 \cdot 79 \cdot 83 \\
307\,583\,652\,416\,968\,363\,617\,090 &= 2 \cdot 3^3 \cdot 5 \cdot 7 \cdot 13 \cdot 19 \cdot 37 \cdot 41 \cdot 53 \cdot 59 \cdot 61 \cdot 67 \cdot 71 \cdot 73 \cdot 79 \cdot 83 \\
410\,111\,536\,555\,957\,818\,156\,120 &= 2^3 \cdot 3^2 \cdot 5 \cdot 7 \cdot 13 \cdot 19 \cdot 37 \cdot 41 \cdot 53 \cdot 59 \cdot 61 \cdot 67 \cdot 71 \cdot 73 \cdot 79 \cdot 83 \\
520\,952\,492\,381\,892\,363\,603\,720 &= 2^3 \cdot 3^2 \cdot 5 \cdot 7 \cdot 13 \cdot 19 \cdot 41 \cdot 47 \cdot 53 \cdot 59 \cdot 61 \cdot 67 \cdot 71 \cdot 73 \cdot 79 \cdot 83 \\
630\,626\,701\,304\,396\,019\,099\,240 &= 2^3 \cdot 3^2 \cdot 5 \cdot 7 \cdot 13 \cdot 23 \cdot 41 \cdot 47 \cdot 53 \cdot 59 \cdot 61 \cdot 67 \cdot 71 \cdot 73 \cdot 79 \cdot 83 \\
727\,646\,193\,812\,764\,637\,422\,200 &= 2^3 \cdot 3^3 \cdot 5^2 \cdot 7 \cdot 23 \cdot 41 \cdot 47 \cdot 53 \cdot 59 \cdot 61 \cdot 67 \cdot 71 \cdot 73 \cdot 79 \cdot 83 \\
800\,410\,813\,194\,041\,101\,164\,420 &= 2^2 \cdot 3^3 \cdot 5 \cdot 7 \cdot 11 \cdot 23 \cdot 41 \cdot 47 \cdot 53 \cdot 59 \cdot 61 \cdot 67 \cdot 71 \cdot 73 \cdot 79 \cdot 83 \\
839\,455\,243\,105\,945\,545\,123\,660 &= 2^2 \cdot 3^3 \cdot 5 \cdot 7 \cdot 11 \cdot 23 \cdot 43 \cdot 47 \cdot 53 \cdot 59 \cdot 61 \cdot 67 \cdot 71 \cdot 73 \cdot 79 \cdot 83
\end{aligned}
$$

<u>Row 84</u>

$$84 = 2^2 \cdot 3 \cdot 7$$
$$3\,486 = 2 \cdot 3 \cdot 7 \cdot 83$$
$$95\,284 = 2^2 \cdot 7 \cdot 41 \cdot 83$$
$$1\,929\,501 = 3^4 \cdot 7 \cdot 41 \cdot 83$$
$$30\,872\,016 = 2^4 \cdot 3^4 \cdot 7 \cdot 41 \cdot 83$$
$$406\,481\,544 = 2^3 \cdot 3^3 \cdot 7 \cdot 41 \cdot 79 \cdot 83$$
$$4\,529\,365\,776 = 2^4 \cdot 3^4 \cdot 13 \cdot 41 \cdot 79 \cdot 83$$
$$43\,595\,145\,594 = 2 \cdot 3^4 \cdot 7 \cdot 11 \cdot 13 \cdot 41 \cdot 79 \cdot 83$$
$$368\,136\,785\,016 = 2^3 \cdot 3^2 \cdot 7 \cdot 11 \cdot 13 \cdot 19 \cdot 41 \cdot 79 \cdot 83$$
$$2\,761\,025\,887\,620 = 2^2 \cdot 3^3 \cdot 5 \cdot 7 \cdot 11 \cdot 13 \cdot 19 \cdot 41 \cdot 79 \cdot 83$$
$$18\,574\,174\,153\,080 = 2^3 \cdot 3^3 \cdot 5 \cdot 7 \cdot 13 \cdot 19 \cdot 37 \cdot 41 \cdot 79 \cdot 83$$
$$112\,992\,892\,764\,570 = 2 \cdot 3^2 \cdot 5 \cdot 7 \cdot 13 \cdot 19 \cdot 37 \cdot 41 \cdot 73 \cdot 79 \cdot 83$$
$$625\,806\,790\,696\,080 = 2^4 \cdot 3^4 \cdot 5 \cdot 7 \cdot 19 \cdot 37 \cdot 41 \cdot 73 \cdot 79 \cdot 83$$
$$3\,173\,734\,438\,530\,120 = 2^3 \cdot 3^4 \cdot 5 \cdot 19 \cdot 37 \cdot 41 \cdot 71 \cdot 73 \cdot 79 \cdot 83$$
$$14\,810\,760\,713\,140\,560 = 2^4 \cdot 3^3 \cdot 5 \cdot 7 \cdot 19 \cdot 37 \cdot 41 \cdot 71 \cdot 73 \cdot 79 \cdot 83$$
$$63\,871\,405\,575\,418\,665 = 3^4 \cdot 5 \cdot 7 \cdot 19 \cdot 23 \cdot 37 \cdot 41 \cdot 71 \cdot 73 \cdot 79 \cdot 83$$
$$255\,485\,622\,301\,674\,660 = 2^2 \cdot 3^4 \cdot 5 \cdot 7 \cdot 19 \cdot 23 \cdot 37 \cdot 41 \cdot 71 \cdot 73 \cdot 79 \cdot 83$$
$$950\,974\,260\,789\,566\,790 = 2 \cdot 3^2 \cdot 5 \cdot 7 \cdot 19 \cdot 23 \cdot 37 \cdot 41 \cdot 67 \cdot 71 \cdot 73 \cdot 79 \cdot 83$$
$$3\,303\,384\,274\,321\,653\,060 = 2^2 \cdot 3^3 \cdot 5 \cdot 7 \cdot 11 \cdot 23 \cdot 37 \cdot 41 \cdot 67 \cdot 71 \cdot 73 \cdot 79 \cdot 83$$
$$10\,735\,998\,891\,545\,372\,445 = 3^3 \cdot 5 \cdot 7 \cdot 11 \cdot 13 \cdot 23 \cdot 37 \cdot 41 \cdot 67 \cdot 71 \cdot 73 \cdot 79 \cdot 83$$
$$32\,719\,234\,717\,090\,658\,880 = 2^6 \cdot 3^2 \cdot 5 \cdot 11 \cdot 13 \cdot 23 \cdot 37 \cdot 41 \cdot 67 \cdot 71 \cdot 73 \cdot 79 \cdot 83$$
$$93\,695\,990\,326\,214\,159\,520 = 2^5 \cdot 3^4 \cdot 5 \cdot 7 \cdot 13 \cdot 23 \cdot 37 \cdot 41 \cdot 67 \cdot 71 \cdot 73 \cdot 79 \cdot 83$$
$$252\,571\,800\,009\,794\,690\,880 = 2^6 \cdot 3^4 \cdot 5 \cdot 7 \cdot 13 \cdot 31 \cdot 37 \cdot 41 \cdot 67 \cdot 71 \cdot 73 \cdot 79 \cdot 83$$
$$641\,953\,325\,024\,894\,839\,320 = 2^3 \cdot 3^3 \cdot 5 \cdot 7 \cdot 13 \cdot 31 \cdot 37 \cdot 41 \cdot 61 \cdot 67 \cdot 71 \cdot 73 \cdot 79 \cdot 83$$
$$1\,540\,687\,980\,059\,747\,614\,368 = 2^5 \cdot 3^4 \cdot 7 \cdot 13 \cdot 31 \cdot 37 \cdot 41 \cdot 61 \cdot 67 \cdot 71 \cdot 73 \cdot 79 \cdot 83$$
$$3\,496\,176\,570\,135\,581\,124\,912 = 2^4 \cdot 3^4 \cdot 7 \cdot 31 \cdot 37 \cdot 41 \cdot 59 \cdot 61 \cdot 67 \cdot 71 \cdot 73 \cdot 79 \cdot 83$$
$$7\,510\,305\,224\,735\,692\,786\,848 = 2^5 \cdot 3 \cdot 7 \cdot 29 \cdot 31 \cdot 37 \cdot 41 \cdot 59 \cdot 61 \cdot 67 \cdot 71 \cdot 73 \cdot 79 \cdot 83$$
$$15\,288\,835\,636\,069\,088\,887\,512 = 2^3 \cdot 3^2 \cdot 19 \cdot 29 \cdot 31 \cdot 37 \cdot 41 \cdot 59 \cdot 61 \cdot 67 \cdot 71 \cdot 73 \cdot 79 \cdot 83$$
$$29\,523\,268\,814\,478\,240\,610\,368 = 2^6 \cdot 3^2 \cdot 7 \cdot 19 \cdot 31 \cdot 37 \cdot 41 \cdot 59 \cdot 61 \cdot 67 \cdot 71 \cdot 73 \cdot 79 \cdot 83$$
$$54\,125\,992\,826\,543\,441\,119\,008 = 2^5 \cdot 3 \cdot 7 \cdot 11 \cdot 19 \cdot 31 \cdot 37 \cdot 41 \cdot 59 \cdot 61 \cdot 67 \cdot 71 \cdot 73 \cdot 79 \cdot 83$$
$$94\,283\,987\,504\,301\,478\,078\,272 = 2^6 \cdot 3^4 \cdot 7 \cdot 11 \cdot 19 \cdot 37 \cdot 41 \cdot 59 \cdot 61 \cdot 67 \cdot 71 \cdot 73 \cdot 79 \cdot 83$$
$$156\,157\,854\,303\,999\,323\,067\,138 = 2 \cdot 3^4 \cdot 7 \cdot 11 \cdot 19 \cdot 37 \cdot 41 \cdot 53 \cdot 59 \cdot 61 \cdot 67 \cdot 71 \cdot 73 \cdot 79 \cdot 83$$
$$246\,066\,921\,933\,574\,690\,893\,672 = 2^3 \cdot 3^3 \cdot 7 \cdot 13 \cdot 19 \cdot 37 \cdot 41 \cdot 53 \cdot 59 \cdot 61 \cdot 67 \cdot 71 \cdot 73 \cdot 79 \cdot 83$$
$$369\,100\,382\,900\,362\,036\,340\,508 = 2^2 \cdot 3^4 \cdot 7 \cdot 13 \cdot 19 \cdot 37 \cdot 41 \cdot 53 \cdot 59 \cdot 61 \cdot 67 \cdot 71 \cdot 73 \cdot 79 \cdot 83$$
$$527\,286\,261\,286\,231\,480\,486\,440 = 2^3 \cdot 3^4 \cdot 5 \cdot 13 \cdot 19 \cdot 37 \cdot 41 \cdot 53 \cdot 59 \cdot 61 \cdot 67 \cdot 71 \cdot 73 \cdot 79 \cdot 83$$
$$717\,695\,188\,972\,926\,181\,773\,210 = 2 \cdot 3^2 \cdot 5 \cdot 7^2 \cdot 13 \cdot 19 \cdot 37 \cdot 41 \cdot 53 \cdot 59 \cdot 61 \cdot 67 \cdot 71 \cdot 73 \cdot 79 \cdot 83$$
$$931\,064\,028\,937\,850\,181\,759\,840 = 2^5 \cdot 3^3 \cdot 5 \cdot 7^2 \cdot 13 \cdot 19 \cdot 41 \cdot 53 \cdot 59 \cdot 61 \cdot 67 \cdot 71 \cdot 73 \cdot 79 \cdot 83$$
$$1\,151\,579\,193\,686\,288\,382\,702\,960 = 2^4 \cdot 3^3 \cdot 5 \cdot 7^2 \cdot 13 \cdot 41 \cdot 47 \cdot 53 \cdot 59 \cdot 61 \cdot 67 \cdot 71 \cdot 73 \cdot 79 \cdot 83$$
$$1\,358\,272\,895\,117\,160\,656\,521\,440 = 2^5 \cdot 3^2 \cdot 5 \cdot 7^2 \cdot 23 \cdot 41 \cdot 47 \cdot 53 \cdot 59 \cdot 61 \cdot 67 \cdot 71 \cdot 73 \cdot 79 \cdot 83$$
$$1\,528\,057\,007\,006\,805\,738\,586\,620 = 2^2 \cdot 3^4 \cdot 5 \cdot 7^2 \cdot 23 \cdot 41 \cdot 47 \cdot 53 \cdot 59 \cdot 61 \cdot 67 \cdot 71 \cdot 73 \cdot 79 \cdot 83$$
$$1\,639\,866\,056\,299\,986\,646\,288\,080 = 2^4 \cdot 3^4 \cdot 5 \cdot 7^2 \cdot 11 \cdot 23 \cdot 47 \cdot 53 \cdot 59 \cdot 61 \cdot 67 \cdot 71 \cdot 73 \cdot 79 \cdot 83$$
$$1\,678\,910\,486\,211\,891\,090\,247\,320 = 2^3 \cdot 3^3 \cdot 5 \cdot 7 \cdot 11 \cdot 23 \cdot 43 \cdot 47 \cdot 53 \cdot 59 \cdot 61 \cdot 67 \cdot 71 \cdot 73 \cdot 79 \cdot 83$$

<u>Row 85</u>

$$85 = 5 \cdot 17$$
$$3\,570 = 2 \cdot 3 \cdot 5 \cdot 7 \cdot 17$$
$$98\,770 = 2 \cdot 5 \cdot 7 \cdot 17 \cdot 83$$
$$2\,024\,785 = 5 \cdot 7 \cdot 17 \cdot 41 \cdot 83$$
$$32\,801\,517 = 3^4 \cdot 7 \cdot 17 \cdot 41 \cdot 83$$
$$437\,353\,560 = 2^3 \cdot 3^3 \cdot 5 \cdot 7 \cdot 17 \cdot 41 \cdot 83$$
$$4\,935\,847\,320 = 2^3 \cdot 3^3 \cdot 5 \cdot 17 \cdot 41 \cdot 79 \cdot 83$$
$$48\,124\,511\,370 = 2 \cdot 3^4 \cdot 5 \cdot 13 \cdot 17 \cdot 41 \cdot 79 \cdot 83$$
$$411\,731\,930\,610 = 2 \cdot 3^2 \cdot 5 \cdot 7 \cdot 11 \cdot 13 \cdot 17 \cdot 41 \cdot 79 \cdot 83$$
$$3\,129\,162\,672\,636 = 2^2 \cdot 3^2 \cdot 7 \cdot 11 \cdot 13 \cdot 17 \cdot 19 \cdot 41 \cdot 79 \cdot 83$$
$$21\,335\,200\,040\,700 = 2^2 \cdot 3^3 \cdot 5^2 \cdot 7 \cdot 13 \cdot 17 \cdot 19 \cdot 41 \cdot 79 \cdot 83$$
$$131\,567\,066\,917\,650 = 2 \cdot 3^2 \cdot 5^2 \cdot 7 \cdot 13 \cdot 17 \cdot 19 \cdot 37 \cdot 41 \cdot 79 \cdot 83$$
$$738\,799\,683\,460\,650 = 2 \cdot 3^2 \cdot 5^2 \cdot 7 \cdot 17 \cdot 19 \cdot 37 \cdot 41 \cdot 73 \cdot 79 \cdot 83$$
$$3\,799\,541\,229\,226\,200 = 2^3 \cdot 3^4 \cdot 5^2 \cdot 17 \cdot 19 \cdot 37 \cdot 41 \cdot 73 \cdot 79 \cdot 83$$
$$17\,984\,495\,151\,670\,680 = 2^3 \cdot 3^3 \cdot 5 \cdot 17 \cdot 19 \cdot 37 \cdot 41 \cdot 71 \cdot 73 \cdot 79 \cdot 83$$
$$78\,682\,166\,288\,559\,225 = 3^3 \cdot 5^2 \cdot 7 \cdot 17 \cdot 19 \cdot 37 \cdot 41 \cdot 71 \cdot 73 \cdot 79 \cdot 83$$
$$319\,357\,027\,877\,093\,325 = 3^4 \cdot 5^2 \cdot 7 \cdot 19 \cdot 23 \cdot 37 \cdot 41 \cdot 71 \cdot 73 \cdot 79 \cdot 83$$
$$1\,206\,459\,883\,091\,241\,450 = 2 \cdot 3^2 \cdot 5^2 \cdot 7 \cdot 17 \cdot 19 \cdot 23 \cdot 37 \cdot 41 \cdot 71 \cdot 73 \cdot 79 \cdot 83$$
$$4\,254\,358\,535\,111\,219\,850 = 2 \cdot 3^2 \cdot 5^2 \cdot 7 \cdot 17 \cdot 23 \cdot 37 \cdot 41 \cdot 67 \cdot 71 \cdot 73 \cdot 79 \cdot 83$$
$$14\,039\,383\,165\,867\,025\,505 = 3^3 \cdot 5 \cdot 7 \cdot 11 \cdot 17 \cdot 23 \cdot 37 \cdot 41 \cdot 67 \cdot 71 \cdot 73 \cdot 79 \cdot 83$$
$$43\,455\,233\,608\,636\,031\,325 = 3^2 \cdot 5^2 \cdot 11 \cdot 13 \cdot 17 \cdot 23 \cdot 37 \cdot 41 \cdot 67 \cdot 71 \cdot 73 \cdot 79 \cdot 83$$
$$126\,415\,225\,043\,304\,818\,400 = 2^5 \cdot 3^2 \cdot 5^2 \cdot 13 \cdot 17 \cdot 23 \cdot 37 \cdot 41 \cdot 67 \cdot 71 \cdot 73 \cdot 79 \cdot 83$$
$$346\,267\,790\,336\,008\,850\,400 = 2^5 \cdot 3^4 \cdot 5^2 \cdot 7 \cdot 13 \cdot 17 \cdot 37 \cdot 41 \cdot 67 \cdot 71 \cdot 73 \cdot 79 \cdot 83$$
$$894\,525\,125\,034\,689\,530\,200 = 2^3 \cdot 3^3 \cdot 5^2 \cdot 7 \cdot 13 \cdot 17 \cdot 31 \cdot 37 \cdot 41 \cdot 67 \cdot 71 \cdot 73 \cdot 79 \cdot 83$$
$$2\,182\,641\,305\,084\,642\,453\,688 = 2^3 \cdot 3^3 \cdot 7 \cdot 13 \cdot 17 \cdot 31 \cdot 37 \cdot 41 \cdot 61 \cdot 67 \cdot 71 \cdot 73 \cdot 79 \cdot 83$$
$$5\,036\,864\,550\,195\,328\,739\,280 = 2^4 \cdot 3^4 \cdot 5 \cdot 7 \cdot 17 \cdot 31 \cdot 37 \cdot 41 \cdot 61 \cdot 67 \cdot 71 \cdot 73 \cdot 79 \cdot 83$$
$$11\,006\,481\,794\,871\,273\,911\,760 = 2^4 \cdot 3 \cdot 5 \cdot 7 \cdot 17 \cdot 31 \cdot 37 \cdot 41 \cdot 59 \cdot 61 \cdot 67 \cdot 71 \cdot 73 \cdot 79 \cdot 83$$
$$22\,799\,140\,860\,804\,781\,674\,360 = 2^3 \cdot 3 \cdot 5 \cdot 17 \cdot 29 \cdot 31 \cdot 37 \cdot 41 \cdot 59 \cdot 61 \cdot 67 \cdot 71 \cdot 73 \cdot 79 \cdot 83$$
$$44\,812\,104\,450\,547\,329\,497\,880 = 2^3 \cdot 3^2 \cdot 5 \cdot 17 \cdot 19 \cdot 31 \cdot 37 \cdot 41 \cdot 59 \cdot 61 \cdot 67 \cdot 71 \cdot 73 \cdot 79 \cdot 83$$
$$83\,649\,261\,641\,021\,681\,729\,376 = 2^5 \cdot 3 \cdot 7 \cdot 17 \cdot 19 \cdot 31 \cdot 37 \cdot 41 \cdot 59 \cdot 61 \cdot 67 \cdot 71 \cdot 73 \cdot 79 \cdot 83$$
$$148\,409\,980\,330\,844\,919\,197\,280 = 2^5 \cdot 3 \cdot 5 \cdot 7 \cdot 11 \cdot 17 \cdot 19 \cdot 37 \cdot 41 \cdot 59 \cdot 61 \cdot 67 \cdot 71 \cdot 73 \cdot 79 \cdot 83$$
$$250\,441\,841\,808\,300\,801\,145\,410 = 2 \cdot 3^4 \cdot 5 \cdot 7 \cdot 11 \cdot 17 \cdot 19 \cdot 37 \cdot 41 \cdot 59 \cdot 61 \cdot 67 \cdot 71 \cdot 73 \cdot 79 \cdot 83$$
$$402\,224\,776\,237\,574\,013\,960\,810 = 2 \cdot 3^3 \cdot 5 \cdot 7 \cdot 17 \cdot 19 \cdot 37 \cdot 41 \cdot 53 \cdot 59 \cdot 61 \cdot 67 \cdot 71 \cdot 73 \cdot 79 \cdot 83$$
$$615\,167\,304\,833\,936\,727\,234\,180 = 2^2 \cdot 3^3 \cdot 5 \cdot 7 \cdot 13 \cdot 19 \cdot 37 \cdot 41 \cdot 53 \cdot 59 \cdot 61 \cdot 67 \cdot 71 \cdot 73 \cdot 79 \cdot 83$$
$$896\,386\,644\,186\,593\,516\,826\,948 = 2^2 \cdot 3^4 \cdot 13 \cdot 17 \cdot 19 \cdot 37 \cdot 41 \cdot 53 \cdot 59 \cdot 61 \cdot 67 \cdot 71 \cdot 73 \cdot 79 \cdot 83$$
$$1\,244\,981\,450\,259\,157\,662\,259\,650 = 2 \cdot 3^2 \cdot 5^2 \cdot 13 \cdot 17 \cdot 19 \cdot 37 \cdot 41 \cdot 53 \cdot 59 \cdot 61 \cdot 67 \cdot 71 \cdot 73 \cdot 79 \cdot 83$$
$$1\,648\,759\,217\,910\,776\,363\,533\,050 = 2 \cdot 3^2 \cdot 5^2 \cdot 7^2 \cdot 13 \cdot 17 \cdot 19 \cdot 41 \cdot 53 \cdot 59 \cdot 61 \cdot 67 \cdot 71 \cdot 73 \cdot 79 \cdot 83$$
$$2\,082\,643\,222\,624\,138\,564\,462\,800 = 2^4 \cdot 3^3 \cdot 5^2 \cdot 7^2 \cdot 13 \cdot 17 \cdot 41 \cdot 53 \cdot 59 \cdot 61 \cdot 67 \cdot 71 \cdot 73 \cdot 79 \cdot 83$$
$$2\,509\,852\,088\,803\,449\,039\,224\,400 = 2^4 \cdot 3^2 \cdot 5^2 \cdot 7^2 \cdot 17 \cdot 41 \cdot 47 \cdot 53 \cdot 59 \cdot 61 \cdot 67 \cdot 71 \cdot 73 \cdot 79 \cdot 83$$
$$2\,886\,329\,902\,123\,966\,395\,108\,060 = 2^2 \cdot 3^2 \cdot 5 \cdot 7^2 \cdot 17 \cdot 23 \cdot 41 \cdot 47 \cdot 53 \cdot 59 \cdot 61 \cdot 67 \cdot 71 \cdot 73 \cdot 79 \cdot 83$$
$$3\,167\,923\,063\,306\,792\,384\,874\,700 = 2^2 \cdot 3^4 \cdot 5^2 \cdot 7^2 \cdot 17 \cdot 23 \cdot 47 \cdot 53 \cdot 59 \cdot 61 \cdot 67 \cdot 71 \cdot 73 \cdot 79 \cdot 83$$
$$3\,318\,776\,542\,511\,877\,736\,535\,400 = 2^3 \cdot 3^3 \cdot 5^2 \cdot 7 \cdot 11 \cdot 17 \cdot 23 \cdot 47 \cdot 53 \cdot 59 \cdot 61 \cdot 67 \cdot 71 \cdot 73 \cdot 79 \cdot 83$$

Row 86

$$86 = 2 \cdot 43$$
$$3\,655 = 5 \cdot 17 \cdot 43$$
$$102\,340 = 2^2 \cdot 5 \cdot 7 \cdot 17 \cdot 43$$
$$2\,123\,555 = 5 \cdot 7 \cdot 17 \cdot 43 \cdot 83$$
$$34\,826\,302 = 2 \cdot 7 \cdot 17 \cdot 41 \cdot 43 \cdot 83$$
$$470\,155\,077 = 3^3 \cdot 7 \cdot 17 \cdot 41 \cdot 43 \cdot 83$$
$$5\,373\,200\,880 = 2^4 \cdot 3^3 \cdot 5 \cdot 17 \cdot 41 \cdot 43 \cdot 83$$
$$53\,060\,358\,690 = 2 \cdot 3^3 \cdot 5 \cdot 17 \cdot 41 \cdot 43 \cdot 79 \cdot 83$$
$$459\,856\,441\,980 = 2^2 \cdot 3^2 \cdot 5 \cdot 13 \cdot 17 \cdot 41 \cdot 43 \cdot 79 \cdot 83$$
$$3\,540\,894\,603\,246 = 2 \cdot 3^2 \cdot 7 \cdot 11 \cdot 13 \cdot 17 \cdot 41 \cdot 43 \cdot 79 \cdot 83$$
$$24\,464\,362\,713\,336 = 2^3 \cdot 3^2 \cdot 7 \cdot 13 \cdot 17 \cdot 19 \cdot 41 \cdot 43 \cdot 79 \cdot 83$$
$$152\,902\,266\,958\,350 = 2 \cdot 3^2 \cdot 5^2 \cdot 7 \cdot 13 \cdot 17 \cdot 19 \cdot 41 \cdot 43 \cdot 79 \cdot 83$$
$$870\,366\,750\,378\,300 = 2^2 \cdot 3^2 \cdot 5^2 \cdot 7 \cdot 17 \cdot 19 \cdot 37 \cdot 41 \cdot 43 \cdot 79 \cdot 83$$
$$4\,538\,340\,912\,686\,850 = 2 \cdot 3^2 \cdot 5^2 \cdot 17 \cdot 19 \cdot 37 \cdot 41 \cdot 43 \cdot 73 \cdot 79 \cdot 83$$
$$21\,784\,036\,380\,896\,880 = 2^4 \cdot 3^3 \cdot 5 \cdot 17 \cdot 19 \cdot 37 \cdot 41 \cdot 43 \cdot 73 \cdot 79 \cdot 83$$
$$96\,666\,661\,440\,229\,905 = 3^3 \cdot 5 \cdot 17 \cdot 19 \cdot 37 \cdot 41 \cdot 43 \cdot 71 \cdot 73 \cdot 79 \cdot 83$$
$$398\,039\,194\,165\,652\,550 = 2 \cdot 3^3 \cdot 5^2 \cdot 7 \cdot 19 \cdot 37 \cdot 41 \cdot 43 \cdot 71 \cdot 73 \cdot 79 \cdot 83$$
$$1\,525\,816\,910\,968\,334\,775 = 3^2 \cdot 5^2 \cdot 7 \cdot 19 \cdot 23 \cdot 37 \cdot 41 \cdot 43 \cdot 71 \cdot 73 \cdot 79 \cdot 83$$
$$5\,460\,818\,418\,202\,461\,300 = 2^2 \cdot 3^2 \cdot 5^2 \cdot 7 \cdot 17 \cdot 23 \cdot 37 \cdot 41 \cdot 43 \cdot 71 \cdot 73 \cdot 79 \cdot 83$$
$$18\,293\,741\,700\,978\,245\,355 = 3^2 \cdot 5 \cdot 7 \cdot 17 \cdot 23 \cdot 37 \cdot 41 \cdot 43 \cdot 67 \cdot 71 \cdot 73 \cdot 79 \cdot 83$$
$$57\,494\,616\,774\,503\,056\,830 = 2 \cdot 3^2 \cdot 5 \cdot 11 \cdot 17 \cdot 23 \cdot 37 \cdot 41 \cdot 43 \cdot 67 \cdot 71 \cdot 73 \cdot 79 \cdot 83$$
$$169\,870\,458\,651\,940\,849\,725 = 3^2 \cdot 5^2 \cdot 13 \cdot 17 \cdot 23 \cdot 37 \cdot 41 \cdot 43 \cdot 67 \cdot 71 \cdot 73 \cdot 79 \cdot 83$$
$$472\,683\,015\,379\,313\,668\,800 = 2^6 \cdot 3^2 \cdot 5^2 \cdot 13 \cdot 17 \cdot 37 \cdot 41 \cdot 43 \cdot 67 \cdot 71 \cdot 73 \cdot 79 \cdot 83$$
$$1\,240\,792\,915\,370\,698\,380\,600 = 2^3 \cdot 3^3 \cdot 5^2 \cdot 7 \cdot 13 \cdot 17 \cdot 37 \cdot 41 \cdot 43 \cdot 67 \cdot 71 \cdot 73 \cdot 79 \cdot 83$$
$$3\,077\,166\,430\,119\,331\,983\,888 = 2^4 \cdot 3^3 \cdot 7 \cdot 13 \cdot 17 \cdot 31 \cdot 37 \cdot 41 \cdot 43 \cdot 67 \cdot 71 \cdot 73 \cdot 79 \cdot 83$$
$$7\,219\,505\,855\,279\,971\,192\,968 = 2^3 \cdot 3^3 \cdot 7 \cdot 17 \cdot 31 \cdot 37 \cdot 41 \cdot 43 \cdot 61 \cdot 67 \cdot 71 \cdot 73 \cdot 79 \cdot 83$$
$$16\,043\,346\,345\,066\,602\,651\,040 = 2^5 \cdot 3 \cdot 5 \cdot 7 \cdot 17 \cdot 31 \cdot 37 \cdot 41 \cdot 43 \cdot 61 \cdot 67 \cdot 71 \cdot 73 \cdot 79 \cdot 83$$
$$33\,805\,622\,655\,676\,055\,586\,120 = 2^3 \cdot 3 \cdot 5 \cdot 17 \cdot 31 \cdot 37 \cdot 41 \cdot 43 \cdot 59 \cdot 61 \cdot 67 \cdot 71 \cdot 73 \cdot 79 \cdot 83$$
$$67\,611\,245\,311\,352\,111\,172\,240 = 2^4 \cdot 3 \cdot 5 \cdot 17 \cdot 31 \cdot 37 \cdot 41 \cdot 43 \cdot 59 \cdot 61 \cdot 67 \cdot 71 \cdot 73 \cdot 79 \cdot 83$$
$$128\,461\,366\,091\,569\,011\,227\,256 = 2^3 \cdot 3 \cdot 17 \cdot 19 \cdot 31 \cdot 37 \cdot 41 \cdot 43 \cdot 59 \cdot 61 \cdot 67 \cdot 71 \cdot 73 \cdot 79 \cdot 83$$
$$232\,059\,241\,971\,866\,600\,926\,656 = 2^6 \cdot 3 \cdot 7 \cdot 17 \cdot 19 \cdot 37 \cdot 41 \cdot 43 \cdot 59 \cdot 61 \cdot 67 \cdot 71 \cdot 73 \cdot 79 \cdot 83$$
$$398\,851\,822\,139\,145\,720\,342\,690 = 2 \cdot 3 \cdot 5 \cdot 7 \cdot 11 \cdot 17 \cdot 19 \cdot 37 \cdot 41 \cdot 43 \cdot 59 \cdot 61 \cdot 67 \cdot 71 \cdot 73 \cdot 79 \cdot 83$$
$$652\,666\,618\,045\,874\,815\,106\,220 = 2^2 \cdot 3^3 \cdot 5 \cdot 7 \cdot 17 \cdot 19 \cdot 37 \cdot 41 \cdot 43 \cdot 59 \cdot 61 \cdot 67 \cdot 71 \cdot 73 \cdot 79 \cdot 83$$
$$1\,017\,392\,081\,071\,510\,741\,194\,990 = 2 \cdot 3^3 \cdot 5 \cdot 7 \cdot 19 \cdot 37 \cdot 41 \cdot 43 \cdot 53 \cdot 59 \cdot 61 \cdot 67 \cdot 71 \cdot 73 \cdot 79 \cdot 83$$
$$1\,511\,553\,949\,020\,530\,244\,061\,128 = 2^3 \cdot 3^3 \cdot 13 \cdot 19 \cdot 37 \cdot 41 \cdot 43 \cdot 53 \cdot 59 \cdot 61 \cdot 67 \cdot 71 \cdot 73 \cdot 79 \cdot 83$$
$$2\,141\,368\,094\,445\,751\,179\,086\,598 = 2 \cdot 3^2 \cdot 13 \cdot 17 \cdot 19 \cdot 37 \cdot 41 \cdot 43 \cdot 53 \cdot 59 \cdot 61 \cdot 67 \cdot 71 \cdot 73 \cdot 79 \cdot 83$$
$$2\,893\,740\,668\,169\,934\,025\,792\,700 = 2^2 \cdot 3^2 \cdot 5^2 \cdot 13 \cdot 17 \cdot 19 \cdot 41 \cdot 43 \cdot 53 \cdot 59 \cdot 61 \cdot 67 \cdot 71 \cdot 73 \cdot 79 \cdot 83$$
$$3\,731\,402\,440\,534\,914\,927\,995\,850 = 2 \cdot 3^2 \cdot 5^2 \cdot 7^2 \cdot 13 \cdot 17 \cdot 41 \cdot 43 \cdot 53 \cdot 59 \cdot 61 \cdot 67 \cdot 71 \cdot 73 \cdot 79 \cdot 83$$
$$4\,592\,495\,311\,427\,587\,603\,687\,200 = 2^5 \cdot 3^2 \cdot 5^2 \cdot 7^2 \cdot 17 \cdot 41 \cdot 43 \cdot 53 \cdot 59 \cdot 61 \cdot 67 \cdot 71 \cdot 73 \cdot 79 \cdot 83$$
$$5\,396\,181\,990\,927\,415\,434\,332\,460 = 2^2 \cdot 3^2 \cdot 5 \cdot 7^2 \cdot 17 \cdot 41 \cdot 43 \cdot 47 \cdot 53 \cdot 59 \cdot 61 \cdot 67 \cdot 71 \cdot 73 \cdot 79 \cdot 83$$
$$6\,054\,252\,965\,430\,758\,779\,982\,760 = 2^3 \cdot 3^2 \cdot 5 \cdot 7^2 \cdot 17 \cdot 23 \cdot 43 \cdot 47 \cdot 53 \cdot 59 \cdot 61 \cdot 67 \cdot 71 \cdot 73 \cdot 79 \cdot 83$$
$$6\,486\,699\,605\,818\,670\,121\,410\,100 = 2^2 \cdot 3^3 \cdot 5^2 \cdot 7 \cdot 17 \cdot 23 \cdot 43 \cdot 47 \cdot 53 \cdot 59 \cdot 61 \cdot 67 \cdot 71 \cdot 73 \cdot 79 \cdot 83$$
$$6\,637\,553\,085\,023\,755\,473\,070\,800 = 2^4 \cdot 3^3 \cdot 5^2 \cdot 7 \cdot 11 \cdot 17 \cdot 23 \cdot 47 \cdot 53 \cdot 59 \cdot 61 \cdot 67 \cdot 71 \cdot 73 \cdot 79 \cdot 83$$

<u>Row 87</u>

$$87 = 3 \cdot 29$$
$$3\,741 = 3 \cdot 29 \cdot 43$$
$$105\,995 = 5 \cdot 17 \cdot 29 \cdot 43$$
$$2\,225\,895 = 3 \cdot 5 \cdot 7 \cdot 17 \cdot 29 \cdot 43$$
$$36\,949\,857 = 3 \cdot 7 \cdot 17 \cdot 29 \cdot 43 \cdot 83$$
$$504\,981\,379 = 7 \cdot 17 \cdot 29 \cdot 41 \cdot 43 \cdot 83$$
$$5\,843\,355\,957 = 3^4 \cdot 17 \cdot 29 \cdot 41 \cdot 43 \cdot 83$$
$$58\,433\,559\,570 = 2 \cdot 3^4 \cdot 5 \cdot 17 \cdot 29 \cdot 41 \cdot 43 \cdot 83$$
$$512\,916\,800\,670 = 2 \cdot 3^2 \cdot 5 \cdot 17 \cdot 29 \cdot 41 \cdot 43 \cdot 79 \cdot 83$$
$$4\,000\,751\,045\,226 = 2 \cdot 3^3 \cdot 13 \cdot 17 \cdot 29 \cdot 41 \cdot 43 \cdot 79 \cdot 83$$
$$28\,005\,257\,316\,582 = 2 \cdot 3^3 \cdot 7 \cdot 13 \cdot 17 \cdot 29 \cdot 41 \cdot 43 \cdot 79 \cdot 83$$
$$177\,366\,629\,671\,686 = 2 \cdot 3^2 \cdot 7 \cdot 13 \cdot 17 \cdot 19 \cdot 29 \cdot 41 \cdot 43 \cdot 79 \cdot 83$$
$$1\,023\,269\,017\,336\,650 = 2 \cdot 3^3 \cdot 5^2 \cdot 7 \cdot 17 \cdot 19 \cdot 29 \cdot 41 \cdot 43 \cdot 79 \cdot 83$$
$$5\,408\,707\,663\,065\,150 = 2 \cdot 3^3 \cdot 5^2 \cdot 17 \cdot 19 \cdot 29 \cdot 37 \cdot 41 \cdot 43 \cdot 79 \cdot 83$$
$$26\,322\,377\,293\,583\,730 = 2 \cdot 3^2 \cdot 5 \cdot 17 \cdot 19 \cdot 29 \cdot 37 \cdot 41 \cdot 43 \cdot 73 \cdot 79 \cdot 83$$
$$118\,450\,697\,821\,126\,785 = 3^4 \cdot 5 \cdot 17 \cdot 19 \cdot 29 \cdot 37 \cdot 41 \cdot 43 \cdot 73 \cdot 79 \cdot 83$$
$$494\,705\,855\,605\,882\,455 = 3^4 \cdot 5 \cdot 19 \cdot 29 \cdot 37 \cdot 41 \cdot 43 \cdot 71 \cdot 73 \cdot 79 \cdot 83$$
$$1\,923\,856\,105\,133\,987\,325 = 3^2 \cdot 5^2 \cdot 7 \cdot 19 \cdot 29 \cdot 37 \cdot 41 \cdot 43 \cdot 71 \cdot 73 \cdot 79 \cdot 83$$
$$6\,986\,635\,329\,170\,796\,075 = 3^3 \cdot 5^2 \cdot 7 \cdot 23 \cdot 29 \cdot 37 \cdot 41 \cdot 43 \cdot 71 \cdot 73 \cdot 79 \cdot 83$$
$$23\,754\,560\,119\,180\,706\,655 = 3^3 \cdot 5 \cdot 7 \cdot 17 \cdot 23 \cdot 29 \cdot 37 \cdot 41 \cdot 43 \cdot 71 \cdot 73 \cdot 79 \cdot 83$$
$$75\,788\,358\,475\,481\,302\,185 = 3^2 \cdot 5 \cdot 17 \cdot 23 \cdot 29 \cdot 37 \cdot 41 \cdot 43 \cdot 67 \cdot 71 \cdot 73 \cdot 79 \cdot 83$$
$$227\,365\,075\,426\,443\,906\,555 = 3^3 \cdot 5 \cdot 17 \cdot 23 \cdot 29 \cdot 37 \cdot 41 \cdot 43 \cdot 67 \cdot 71 \cdot 73 \cdot 79 \cdot 83$$
$$642\,553\,474\,031\,254\,518\,525 = 3^3 \cdot 5^2 \cdot 13 \cdot 17 \cdot 29 \cdot 37 \cdot 41 \cdot 43 \cdot 67 \cdot 71 \cdot 73 \cdot 79 \cdot 83$$
$$1\,713\,475\,930\,750\,012\,049\,400 = 2^3 \cdot 3^2 \cdot 5^2 \cdot 13 \cdot 17 \cdot 29 \cdot 37 \cdot 41 \cdot 43 \cdot 67 \cdot 71 \cdot 73 \cdot 79 \cdot 83$$
$$4\,317\,959\,345\,490\,030\,364\,488 = 2^3 \cdot 3^4 \cdot 7 \cdot 13 \cdot 17 \cdot 29 \cdot 37 \cdot 41 \cdot 43 \cdot 67 \cdot 71 \cdot 73 \cdot 79 \cdot 83$$
$$10\,296\,672\,285\,399\,303\,176\,856 = 2^3 \cdot 3^4 \cdot 7 \cdot 17 \cdot 29 \cdot 31 \cdot 37 \cdot 41 \cdot 43 \cdot 67 \cdot 71 \cdot 73 \cdot 79 \cdot 83$$
$$23\,262\,852\,200\,346\,573\,844\,008 = 2^3 \cdot 3 \cdot 7 \cdot 17 \cdot 29 \cdot 31 \cdot 37 \cdot 41 \cdot 43 \cdot 61 \cdot 67 \cdot 71 \cdot 73 \cdot 79 \cdot 83$$
$$49\,848\,969\,000\,742\,658\,237\,160 = 2^3 \cdot 3^2 \cdot 5 \cdot 17 \cdot 29 \cdot 31 \cdot 37 \cdot 41 \cdot 43 \cdot 61 \cdot 67 \cdot 71 \cdot 73 \cdot 79 \cdot 83$$
$$101\,416\,867\,967\,028\,166\,758\,360 = 2^3 \cdot 3^2 \cdot 5 \cdot 17 \cdot 31 \cdot 37 \cdot 41 \cdot 43 \cdot 59 \cdot 61 \cdot 67 \cdot 71 \cdot 73 \cdot 79 \cdot 83$$
$$196\,072\,611\,402\,921\,122\,399\,496 = 2^3 \cdot 3 \cdot 17 \cdot 29 \cdot 31 \cdot 37 \cdot 41 \cdot 43 \cdot 59 \cdot 61 \cdot 67 \cdot 71 \cdot 73 \cdot 79 \cdot 83$$
$$360\,520\,608\,063\,435\,612\,153\,912 = 2^3 \cdot 3^2 \cdot 17 \cdot 19 \cdot 29 \cdot 37 \cdot 41 \cdot 43 \cdot 59 \cdot 61 \cdot 67 \cdot 71 \cdot 73 \cdot 79 \cdot 83$$
$$630\,911\,064\,111\,012\,321\,269\,346 = 2 \cdot 3^2 \cdot 7 \cdot 17 \cdot 19 \cdot 29 \cdot 37 \cdot 41 \cdot 43 \cdot 59 \cdot 61 \cdot 67 \cdot 71 \cdot 73 \cdot 79 \cdot 83$$
$$1\,051\,518\,440\,185\,020\,535\,448\,910 = 2 \cdot 3 \cdot 5 \cdot 7 \cdot 17 \cdot 19 \cdot 29 \cdot 37 \cdot 41 \cdot 43 \cdot 59 \cdot 61 \cdot 67 \cdot 71 \cdot 73 \cdot 79 \cdot 83$$
$$1\,670\,058\,699\,117\,385\,556\,301\,210 = 2 \cdot 3^4 \cdot 5 \cdot 7 \cdot 19 \cdot 29 \cdot 37 \cdot 41 \cdot 43 \cdot 59 \cdot 61 \cdot 67 \cdot 71 \cdot 73 \cdot 79 \cdot 83$$
$$2\,528\,946\,030\,092\,040\,985\,256\,118 = 2 \cdot 3^4 \cdot 19 \cdot 29 \cdot 37 \cdot 41 \cdot 43 \cdot 53 \cdot 59 \cdot 61 \cdot 67 \cdot 71 \cdot 73 \cdot 79 \cdot 83$$
$$3\,652\,922\,043\,466\,281\,423\,147\,726 = 2 \cdot 3^2 \cdot 13 \cdot 19 \cdot 29 \cdot 37 \cdot 41 \cdot 43 \cdot 53 \cdot 59 \cdot 61 \cdot 67 \cdot 71 \cdot 73 \cdot 79 \cdot 83$$
$$5\,035\,108\,762\,615\,685\,204\,879\,298 = 2 \cdot 3^3 \cdot 13 \cdot 17 \cdot 19 \cdot 29 \cdot 41 \cdot 43 \cdot 53 \cdot 59 \cdot 61 \cdot 67 \cdot 71 \cdot 73 \cdot 79 \cdot 83$$
$$6\,625\,143\,108\,704\,848\,953\,788\,550 = 2 \cdot 3^3 \cdot 5^2 \cdot 13 \cdot 17 \cdot 29 \cdot 41 \cdot 43 \cdot 53 \cdot 59 \cdot 61 \cdot 67 \cdot 71 \cdot 73 \cdot 79 \cdot 83$$
$$8\,323\,897\,751\,962\,502\,531\,683\,050 = 2 \cdot 3^2 \cdot 5^2 \cdot 7^2 \cdot 17 \cdot 29 \cdot 41 \cdot 43 \cdot 53 \cdot 59 \cdot 61 \cdot 67 \cdot 71 \cdot 73 \cdot 79 \cdot 83$$
$$9\,988\,677\,302\,355\,003\,038\,019\,660 = 2^2 \cdot 3^3 \cdot 5 \cdot 7^2 \cdot 17 \cdot 29 \cdot 41 \cdot 43 \cdot 53 \cdot 59 \cdot 61 \cdot 67 \cdot 71 \cdot 73 \cdot 79 \cdot 83$$
$$11\,450\,434\,956\,358\,174\,214\,315\,220 = 2^2 \cdot 3^3 \cdot 5 \cdot 7^2 \cdot 17 \cdot 29 \cdot 43 \cdot 47 \cdot 53 \cdot 59 \cdot 61 \cdot 67 \cdot 71 \cdot 73 \cdot 79 \cdot 83$$
$$12\,540\,952\,571\,249\,428\,901\,392\,860 = 2^2 \cdot 3^2 \cdot 5 \cdot 7 \cdot 17 \cdot 23 \cdot 29 \cdot 43 \cdot 47 \cdot 53 \cdot 59 \cdot 61 \cdot 67 \cdot 71 \cdot 73 \cdot 79 \cdot 83$$
$$13\,124\,252\,690\,842\,425\,594\,480\,900 = 2^2 \cdot 3^4 \cdot 5^2 \cdot 7 \cdot 17 \cdot 23 \cdot 29 \cdot 47 \cdot 53 \cdot 59 \cdot 61 \cdot 67 \cdot 71 \cdot 73 \cdot 79 \cdot 83$$

<u>Row 88</u>

$$88 = 2^3 \cdot 11$$
$$3\ 828 = 2^2 \cdot 3 \cdot 11 \cdot 29$$
$$109\ 736 = 2^3 \cdot 11 \cdot 29 \cdot 43$$
$$2\ 331\ 890 = 2 \cdot 5 \cdot 11 \cdot 17 \cdot 29 \cdot 43$$
$$39\ 175\ 752 = 2^3 \cdot 3 \cdot 7 \cdot 11 \cdot 17 \cdot 29 \cdot 43$$
$$541\ 931\ 236 = 2^2 \cdot 7 \cdot 11 \cdot 17 \cdot 29 \cdot 43 \cdot 83$$
$$6\ 348\ 337\ 336 = 2^3 \cdot 11 \cdot 17 \cdot 29 \cdot 41 \cdot 43 \cdot 83$$
$$64\ 276\ 915\ 527 = 3^4 \cdot 11 \cdot 17 \cdot 29 \cdot 41 \cdot 43 \cdot 83$$
$$571\ 350\ 360\ 240 = 2^4 \cdot 3^2 \cdot 5 \cdot 11 \cdot 17 \cdot 29 \cdot 41 \cdot 43 \cdot 83$$
$$4\ 513\ 667\ 845\ 896 = 2^3 \cdot 3^2 \cdot 11 \cdot 17 \cdot 29 \cdot 41 \cdot 43 \cdot 79 \cdot 83$$
$$32\ 006\ 008\ 361\ 808 = 2^4 \cdot 3^3 \cdot 13 \cdot 17 \cdot 29 \cdot 41 \cdot 43 \cdot 79 \cdot 83$$
$$205\ 371\ 886\ 988\ 268 = 2^2 \cdot 3^2 \cdot 7 \cdot 11 \cdot 13 \cdot 17 \cdot 29 \cdot 41 \cdot 43 \cdot 79 \cdot 83$$
$$1\ 200\ 635\ 647\ 008\ 336 = 2^4 \cdot 3^2 \cdot 7 \cdot 11 \cdot 17 \cdot 19 \cdot 29 \cdot 41 \cdot 43 \cdot 79 \cdot 83$$
$$6\ 431\ 976\ 680\ 401\ 800 = 2^3 \cdot 3^3 \cdot 5^2 \cdot 11 \cdot 17 \cdot 19 \cdot 29 \cdot 41 \cdot 43 \cdot 79 \cdot 83$$
$$31\ 731\ 084\ 956\ 648\ 880 = 2^4 \cdot 3^2 \cdot 5 \cdot 11 \cdot 17 \cdot 19 \cdot 29 \cdot 37 \cdot 41 \cdot 43 \cdot 79 \cdot 83$$
$$144\ 773\ 075\ 114\ 710\ 515 = 3^2 \cdot 5 \cdot 11 \cdot 17 \cdot 19 \cdot 29 \cdot 37 \cdot 41 \cdot 43 \cdot 73 \cdot 79 \cdot 83$$
$$613\ 156\ 553\ 427\ 009\ 240 = 2^3 \cdot 3^4 \cdot 5 \cdot 11 \cdot 19 \cdot 29 \cdot 37 \cdot 41 \cdot 43 \cdot 73 \cdot 79 \cdot 83$$
$$2\ 418\ 561\ 960\ 739\ 869\ 780 = 2^2 \cdot 3^2 \cdot 5 \cdot 11 \cdot 19 \cdot 29 \cdot 37 \cdot 41 \cdot 43 \cdot 71 \cdot 73 \cdot 79 \cdot 83$$
$$8\ 910\ 491\ 434\ 304\ 783\ 400 = 2^3 \cdot 3^2 \cdot 5^2 \cdot 7 \cdot 11 \cdot 29 \cdot 37 \cdot 41 \cdot 43 \cdot 71 \cdot 73 \cdot 79 \cdot 83$$
$$30\ 741\ 195\ 448\ 351\ 502\ 730 = 2 \cdot 3^3 \cdot 5 \cdot 7 \cdot 11 \cdot 23 \cdot 29 \cdot 37 \cdot 41 \cdot 43 \cdot 71 \cdot 73 \cdot 79 \cdot 83$$
$$99\ 542\ 918\ 594\ 662\ 008\ 840 = 2^3 \cdot 3^2 \cdot 5 \cdot 11 \cdot 17 \cdot 23 \cdot 29 \cdot 37 \cdot 41 \cdot 43 \cdot 71 \cdot 73 \cdot 79 \cdot 83$$
$$303\ 153\ 433\ 901\ 925\ 208\ 740 = 2^2 \cdot 3^2 \cdot 5 \cdot 17 \cdot 23 \cdot 29 \cdot 37 \cdot 41 \cdot 43 \cdot 67 \cdot 71 \cdot 73 \cdot 79 \cdot 83$$
$$869\ 918\ 549\ 457\ 698\ 425\ 080 = 2^3 \cdot 3^3 \cdot 5 \cdot 11 \cdot 17 \cdot 29 \cdot 37 \cdot 41 \cdot 43 \cdot 67 \cdot 71 \cdot 73 \cdot 79 \cdot 83$$
$$2\ 356\ 029\ 404\ 781\ 266\ 567\ 925 = 3^2 \cdot 5^2 \cdot 11 \cdot 13 \cdot 17 \cdot 29 \cdot 37 \cdot 41 \cdot 43 \cdot 67 \cdot 71 \cdot 73 \cdot 79 \cdot 83$$
$$6\ 031\ 435\ 276\ 240\ 042\ 413\ 888 = 2^6 \cdot 3^2 \cdot 11 \cdot 13 \cdot 17 \cdot 29 \cdot 37 \cdot 41 \cdot 43 \cdot 67 \cdot 71 \cdot 73 \cdot 79 \cdot 83$$
$$14\ 614\ 631\ 630\ 889\ 333\ 541\ 344 = 2^5 \cdot 3^4 \cdot 7 \cdot 11 \cdot 17 \cdot 29 \cdot 37 \cdot 41 \cdot 43 \cdot 67 \cdot 71 \cdot 73 \cdot 79 \cdot 83$$
$$33\ 559\ 524\ 485\ 745\ 877\ 020\ 864 = 2^6 \cdot 3 \cdot 7 \cdot 11 \cdot 17 \cdot 29 \cdot 31 \cdot 37 \cdot 41 \cdot 43 \cdot 67 \cdot 71 \cdot 73 \cdot 79 \cdot 83$$
$$73\ 111\ 821\ 201\ 089\ 232\ 081\ 168 = 2^4 \cdot 3 \cdot 11 \cdot 17 \cdot 29 \cdot 31 \cdot 37 \cdot 41 \cdot 43 \cdot 61 \cdot 67 \cdot 71 \cdot 73 \cdot 79 \cdot 83$$
$$151\ 265\ 836\ 967\ 770\ 824\ 995\ 520 = 2^6 \cdot 3^2 \cdot 5 \cdot 11 \cdot 17 \cdot 31 \cdot 37 \cdot 41 \cdot 43 \cdot 61 \cdot 67 \cdot 71 \cdot 73 \cdot 79 \cdot 83$$
$$297\ 489\ 479\ 369\ 949\ 289\ 157\ 856 = 2^5 \cdot 3 \cdot 11 \cdot 17 \cdot 31 \cdot 37 \cdot 41 \cdot 43 \cdot 59 \cdot 61 \cdot 67 \cdot 71 \cdot 73 \cdot 79 \cdot 83$$
$$556\ 593\ 219\ 466\ 356\ 734\ 553\ 408 = 2^6 \cdot 3 \cdot 11 \cdot 17 \cdot 29 \cdot 37 \cdot 41 \cdot 43 \cdot 59 \cdot 61 \cdot 67 \cdot 71 \cdot 73 \cdot 79 \cdot 83$$
$$991\ 431\ 672\ 174\ 447\ 933\ 423\ 258 = 2 \cdot 3^2 \cdot 11 \cdot 17 \cdot 19 \cdot 29 \cdot 37 \cdot 41 \cdot 43 \cdot 59 \cdot 61 \cdot 67 \cdot 71 \cdot 73 \cdot 79 \cdot 83$$
$$1\ 682\ 429\ 504\ 296\ 032\ 856\ 718\ 256 = 2^4 \cdot 3 \cdot 7 \cdot 17 \cdot 19 \cdot 29 \cdot 37 \cdot 41 \cdot 43 \cdot 59 \cdot 61 \cdot 67 \cdot 71 \cdot 73 \cdot 79 \cdot 83$$
$$2\ 721\ 577\ 139\ 302\ 406\ 091\ 750\ 120 = 2^3 \cdot 3 \cdot 5 \cdot 7 \cdot 11 \cdot 19 \cdot 29 \cdot 37 \cdot 41 \cdot 43 \cdot 59 \cdot 61 \cdot 67 \cdot 71 \cdot 73 \cdot 79 \cdot 83$$
$$4\ 199\ 004\ 729\ 209\ 426\ 541\ 557\ 328 = 2^4 \cdot 3^4 \cdot 11 \cdot 19 \cdot 29 \cdot 37 \cdot 41 \cdot 43 \cdot 59 \cdot 61 \cdot 67 \cdot 71 \cdot 73 \cdot 79 \cdot 83$$
$$6\ 181\ 868\ 073\ 558\ 322\ 408\ 403\ 844 = 2^2 \cdot 3^2 \cdot 11 \cdot 19 \cdot 29 \cdot 37 \cdot 41 \cdot 43 \cdot 53 \cdot 59 \cdot 61 \cdot 67 \cdot 71 \cdot 73 \cdot 79 \cdot 83$$
$$8\ 688\ 030\ 806\ 081\ 966\ 628\ 027\ 024 = 2^4 \cdot 3^2 \cdot 11 \cdot 13 \cdot 19 \cdot 29 \cdot 41 \cdot 43 \cdot 53 \cdot 59 \cdot 61 \cdot 67 \cdot 71 \cdot 73 \cdot 79 \cdot 83$$
$$11\ 660\ 251\ 871\ 320\ 534\ 158\ 667\ 848 = 2^3 \cdot 3^3 \cdot 11 \cdot 13 \cdot 17 \cdot 29 \cdot 41 \cdot 43 \cdot 53 \cdot 59 \cdot 61 \cdot 67 \cdot 71 \cdot 73 \cdot 79 \cdot 83$$
$$14\ 949\ 040\ 860\ 667\ 351\ 485\ 471\ 600 = 2^4 \cdot 3^2 \cdot 5^2 \cdot 11 \cdot 17 \cdot 29 \cdot 41 \cdot 43 \cdot 53 \cdot 59 \cdot 61 \cdot 67 \cdot 71 \cdot 73 \cdot 79 \cdot 83$$
$$18\ 312\ 575\ 054\ 317\ 505\ 569\ 702\ 710 = 2 \cdot 3^2 \cdot 5 \cdot 7^2 \cdot 11 \cdot 17 \cdot 29 \cdot 41 \cdot 43 \cdot 53 \cdot 59 \cdot 61 \cdot 67 \cdot 71 \cdot 73 \cdot 79 \cdot 83$$
$$21\ 439\ 112\ 258\ 713\ 177\ 252\ 334\ 880 = 2^5 \cdot 3^3 \cdot 5 \cdot 7^2 \cdot 11 \cdot 17 \cdot 29 \cdot 43 \cdot 53 \cdot 59 \cdot 61 \cdot 67 \cdot 71 \cdot 73 \cdot 79 \cdot 83$$
$$23\ 991\ 387\ 527\ 607\ 603\ 115\ 708\ 080 = 2^4 \cdot 3^2 \cdot 5 \cdot 7 \cdot 11 \cdot 17 \cdot 29 \cdot 43 \cdot 47 \cdot 53 \cdot 59 \cdot 61 \cdot 67 \cdot 71 \cdot 73 \cdot 79 \cdot 83$$
$$25\ 665\ 205\ 262\ 091\ 854\ 495\ 873\ 760 = 2^5 \cdot 3^2 \cdot 5 \cdot 7 \cdot 11 \cdot 17 \cdot 23 \cdot 29 \cdot 47 \cdot 53 \cdot 59 \cdot 61 \cdot 67 \cdot 71 \cdot 73 \cdot 79 \cdot 83$$
$$26\ 248\ 505\ 381\ 684\ 851\ 188\ 961\ 800 = 2^3 \cdot 3^4 \cdot 5^2 \cdot 7 \cdot 17 \cdot 23 \cdot 29 \cdot 47 \cdot 53 \cdot 59 \cdot 61 \cdot 67 \cdot 71 \cdot 73 \cdot 79 \cdot 83$$

<u>Row 89</u>

$$89 \text{ is Prime}$$

$$3\,916 = 2^2 \cdot 11 \cdot 89$$

$$113\,564 = 2^2 \cdot 11 \cdot 29 \cdot 89$$

$$2\,441\,626 = 2 \cdot 11 \cdot 29 \cdot 43 \cdot 89$$

$$41\,507\,642 = 2 \cdot 11 \cdot 17 \cdot 29 \cdot 43 \cdot 89$$

$$581\,106\,988 = 2^2 \cdot 7 \cdot 11 \cdot 17 \cdot 29 \cdot 43 \cdot 89$$

$$6\,890\,268\,572 = 2^2 \cdot 11 \cdot 17 \cdot 29 \cdot 43 \cdot 83 \cdot 89$$

$$70\,625\,252\,863 = 11 \cdot 17 \cdot 29 \cdot 41 \cdot 43 \cdot 83 \cdot 89$$

$$635\,627\,275\,767 = 3^2 \cdot 11 \cdot 17 \cdot 29 \cdot 41 \cdot 43 \cdot 83 \cdot 89$$

$$5\,085\,018\,206\,136 = 2^3 \cdot 3^2 \cdot 11 \cdot 17 \cdot 29 \cdot 41 \cdot 43 \cdot 83 \cdot 89$$

$$36\,519\,676\,207\,704 = 2^3 \cdot 3^2 \cdot 17 \cdot 29 \cdot 41 \cdot 43 \cdot 79 \cdot 83 \cdot 89$$

$$237\,377\,895\,350\,076 = 2^2 \cdot 3^2 \cdot 13 \cdot 17 \cdot 29 \cdot 41 \cdot 43 \cdot 79 \cdot 83 \cdot 89$$

$$1\,406\,007\,533\,996\,604 = 2^2 \cdot 3^2 \cdot 7 \cdot 11 \cdot 17 \cdot 29 \cdot 41 \cdot 43 \cdot 79 \cdot 83 \cdot 89$$

$$7\,632\,612\,327\,410\,136 = 2^3 \cdot 3^2 \cdot 11 \cdot 17 \cdot 19 \cdot 29 \cdot 41 \cdot 43 \cdot 79 \cdot 83 \cdot 89$$

$$38\,163\,061\,637\,050\,680 = 2^3 \cdot 3^2 \cdot 5 \cdot 11 \cdot 17 \cdot 19 \cdot 29 \cdot 41 \cdot 43 \cdot 79 \cdot 83 \cdot 89$$

$$176\,504\,160\,071\,359\,395 = 3^2 \cdot 5 \cdot 11 \cdot 17 \cdot 19 \cdot 29 \cdot 37 \cdot 41 \cdot 43 \cdot 79 \cdot 83 \cdot 89$$

$$757\,929\,628\,541\,719\,755 = 3^2 \cdot 5 \cdot 11 \cdot 19 \cdot 29 \cdot 37 \cdot 41 \cdot 43 \cdot 73 \cdot 79 \cdot 83 \cdot 89$$

$$3\,031\,718\,514\,166\,879\,020 = 2^2 \cdot 3^2 \cdot 5 \cdot 11 \cdot 19 \cdot 29 \cdot 37 \cdot 41 \cdot 43 \cdot 73 \cdot 79 \cdot 83 \cdot 89$$

$$11\,329\,053\,395\,044\,653\,180 = 2^2 \cdot 3^2 \cdot 5 \cdot 11 \cdot 29 \cdot 37 \cdot 41 \cdot 43 \cdot 71 \cdot 73 \cdot 79 \cdot 83 \cdot 89$$

$$39\,651\,686\,882\,656\,286\,130 = 2 \cdot 3^2 \cdot 5 \cdot 7 \cdot 11 \cdot 29 \cdot 37 \cdot 41 \cdot 43 \cdot 71 \cdot 73 \cdot 79 \cdot 83 \cdot 89$$

$$130\,284\,114\,043\,013\,511\,570 = 2 \cdot 3^2 \cdot 5 \cdot 11 \cdot 23 \cdot 29 \cdot 37 \cdot 41 \cdot 43 \cdot 71 \cdot 73 \cdot 79 \cdot 83 \cdot 89$$

$$402\,696\,352\,496\,587\,217\,580 = 2^2 \cdot 3^2 \cdot 5 \cdot 17 \cdot 23 \cdot 29 \cdot 37 \cdot 41 \cdot 43 \cdot 71 \cdot 73 \cdot 79 \cdot 83 \cdot 89$$

$$1\,173\,071\,983\,359\,623\,633\,820 = 2^2 \cdot 3^2 \cdot 5 \cdot 17 \cdot 29 \cdot 37 \cdot 41 \cdot 43 \cdot 67 \cdot 71 \cdot 73 \cdot 79 \cdot 83 \cdot 89$$

$$3\,225\,947\,954\,238\,964\,993\,005 = 3^2 \cdot 5 \cdot 11 \cdot 17 \cdot 29 \cdot 37 \cdot 41 \cdot 43 \cdot 67 \cdot 71 \cdot 73 \cdot 79 \cdot 83 \cdot 89$$

$$8\,387\,464\,681\,021\,308\,981\,813 = 3^2 \cdot 11 \cdot 13 \cdot 17 \cdot 29 \cdot 37 \cdot 41 \cdot 43 \cdot 67 \cdot 71 \cdot 73 \cdot 79 \cdot 83 \cdot 89$$

$$20\,646\,066\,907\,129\,375\,955\,232 = 2^5 \cdot 3^2 \cdot 11 \cdot 17 \cdot 29 \cdot 37 \cdot 41 \cdot 43 \cdot 67 \cdot 71 \cdot 73 \cdot 79 \cdot 83 \cdot 89$$

$$48\,174\,156\,116\,635\,210\,562\,208 = 2^5 \cdot 3 \cdot 7 \cdot 11 \cdot 17 \cdot 29 \cdot 37 \cdot 41 \cdot 43 \cdot 67 \cdot 71 \cdot 73 \cdot 79 \cdot 83 \cdot 89$$

$$106\,671\,345\,686\,835\,109\,102\,032 = 2^4 \cdot 3 \cdot 11 \cdot 17 \cdot 29 \cdot 31 \cdot 37 \cdot 41 \cdot 43 \cdot 67 \cdot 71 \cdot 73 \cdot 79 \cdot 83 \cdot 89$$

$$224\,377\,658\,168\,860\,057\,076\,688 = 2^4 \cdot 3 \cdot 11 \cdot 17 \cdot 31 \cdot 37 \cdot 41 \cdot 43 \cdot 61 \cdot 67 \cdot 71 \cdot 73 \cdot 79 \cdot 83 \cdot 89$$

$$448\,755\,316\,337\,720\,114\,153\,376 = 2^5 \cdot 3 \cdot 11 \cdot 17 \cdot 31 \cdot 37 \cdot 41 \cdot 43 \cdot 61 \cdot 67 \cdot 71 \cdot 73 \cdot 79 \cdot 83 \cdot 89$$

$$854\,082\,698\,836\,306\,023\,711\,264 = 2^5 \cdot 3 \cdot 11 \cdot 17 \cdot 37 \cdot 41 \cdot 43 \cdot 59 \cdot 61 \cdot 67 \cdot 71 \cdot 73 \cdot 79 \cdot 83 \cdot 89$$

$$1\,548\,024\,891\,640\,804\,667\,976\,666 = 2 \cdot 3 \cdot 11 \cdot 17 \cdot 29 \cdot 37 \cdot 41 \cdot 43 \cdot 59 \cdot 61 \cdot 67 \cdot 71 \cdot 73 \cdot 79 \cdot 83 \cdot 89$$

$$2\,673\,861\,176\,470\,480\,790\,141\,514 = 2 \cdot 3 \cdot 17 \cdot 19 \cdot 29 \cdot 37 \cdot 41 \cdot 43 \cdot 59 \cdot 61 \cdot 67 \cdot 71 \cdot 73 \cdot 79 \cdot 83 \cdot 89$$

$$4\,404\,006\,643\,598\,438\,948\,468\,376 = 2^3 \cdot 3 \cdot 7 \cdot 19 \cdot 29 \cdot 37 \cdot 41 \cdot 43 \cdot 59 \cdot 61 \cdot 67 \cdot 71 \cdot 73 \cdot 79 \cdot 83 \cdot 89$$

$$6\,920\,581\,868\,511\,832\,633\,307\,448 = 2^3 \cdot 3 \cdot 11 \cdot 19 \cdot 29 \cdot 37 \cdot 41 \cdot 43 \cdot 59 \cdot 61 \cdot 67 \cdot 71 \cdot 73 \cdot 79 \cdot 83 \cdot 89$$

$$10\,380\,872\,802\,767\,748\,949\,961\,172 = 2^2 \cdot 3^2 \cdot 11 \cdot 19 \cdot 29 \cdot 37 \cdot 41 \cdot 43 \cdot 59 \cdot 61 \cdot 67 \cdot 71 \cdot 73 \cdot 79 \cdot 83 \cdot 89$$

$$14\,869\,898\,879\,640\,289\,036\,430\,868 = 2^2 \cdot 3^2 \cdot 11 \cdot 19 \cdot 29 \cdot 41 \cdot 43 \cdot 53 \cdot 59 \cdot 61 \cdot 67 \cdot 71 \cdot 73 \cdot 79 \cdot 83 \cdot 89$$

$$20\,348\,282\,677\,402\,500\,786\,694\,872 = 2^3 \cdot 3^2 \cdot 11 \cdot 13 \cdot 29 \cdot 41 \cdot 43 \cdot 53 \cdot 59 \cdot 61 \cdot 67 \cdot 71 \cdot 73 \cdot 79 \cdot 83 \cdot 89$$

$$26\,609\,292\,731\,987\,885\,644\,139\,448 = 2^3 \cdot 3^2 \cdot 11 \cdot 17 \cdot 29 \cdot 41 \cdot 43 \cdot 53 \cdot 59 \cdot 61 \cdot 67 \cdot 71 \cdot 73 \cdot 79 \cdot 83 \cdot 89$$

$$33\,261\,615\,914\,984\,857\,055\,174\,310 = 2 \cdot 3^2 \cdot 5 \cdot 11 \cdot 17 \cdot 29 \cdot 41 \cdot 43 \cdot 53 \cdot 59 \cdot 61 \cdot 67 \cdot 71 \cdot 73 \cdot 79 \cdot 83 \cdot 89$$

$$39\,751\,687\,313\,030\,682\,822\,037\,590 = 2 \cdot 3^2 \cdot 5 \cdot 7^2 \cdot 11 \cdot 17 \cdot 29 \cdot 43 \cdot 53 \cdot 59 \cdot 61 \cdot 67 \cdot 71 \cdot 73 \cdot 79 \cdot 83 \cdot 89$$

$$45\,430\,499\,786\,320\,780\,368\,042\,960 = 2^4 \cdot 3^2 \cdot 5 \cdot 7 \cdot 11 \cdot 17 \cdot 29 \cdot 43 \cdot 53 \cdot 59 \cdot 61 \cdot 67 \cdot 71 \cdot 73 \cdot 79 \cdot 83 \cdot 89$$

$$49\,656\,592\,789\,699\,457\,611\,581\,840 = 2^4 \cdot 3^2 \cdot 5 \cdot 7 \cdot 11 \cdot 17 \cdot 29 \cdot 47 \cdot 53 \cdot 59 \cdot 61 \cdot 67 \cdot 71 \cdot 73 \cdot 79 \cdot 83 \cdot 89$$

$$51\,913\,710\,643\,776\,705\,684\,835\,560 = 2^3 \cdot 3^2 \cdot 5 \cdot 7 \cdot 17 \cdot 23 \cdot 29 \cdot 47 \cdot 53 \cdot 59 \cdot 61 \cdot 67 \cdot 71 \cdot 73 \cdot 79 \cdot 83 \cdot 89$$

Pascal's Triangle — Prime Factorization — To Center Number (omitting 1's)

Row 90

$$90 = 2 \cdot 3^2 \cdot 5$$
$$4\,005 = 3^2 \cdot 5 \cdot 89$$
$$117\,480 = 2^3 \cdot 3 \cdot 5 \cdot 11 \cdot 89$$
$$2\,555\,190 = 2 \cdot 3^2 \cdot 5 \cdot 11 \cdot 29 \cdot 89$$
$$43\,949\,268 = 2^2 \cdot 3^2 \cdot 11 \cdot 29 \cdot 43 \cdot 89$$
$$622\,614\,630 = 2 \cdot 3 \cdot 5 \cdot 11 \cdot 17 \cdot 29 \cdot 43 \cdot 89$$
$$7\,471\,375\,560 = 2^3 \cdot 3^2 \cdot 5 \cdot 11 \cdot 17 \cdot 29 \cdot 43 \cdot 89$$
$$77\,515\,521\,435 = 3^2 \cdot 5 \cdot 11 \cdot 17 \cdot 29 \cdot 43 \cdot 83 \cdot 89$$
$$706\,252\,528\,630 = 2 \cdot 5 \cdot 11 \cdot 17 \cdot 29 \cdot 41 \cdot 43 \cdot 83 \cdot 89$$
$$5\,720\,645\,481\,903 = 3^4 \cdot 11 \cdot 17 \cdot 29 \cdot 41 \cdot 43 \cdot 83 \cdot 89$$
$$41\,604\,694\,413\,840 = 2^4 \cdot 3^4 \cdot 5 \cdot 17 \cdot 29 \cdot 41 \cdot 43 \cdot 83 \cdot 89$$
$$273\,897\,571\,557\,780 = 2^2 \cdot 3^3 \cdot 5 \cdot 17 \cdot 29 \cdot 41 \cdot 43 \cdot 79 \cdot 83 \cdot 89$$
$$1\,643\,385\,429\,346\,680 = 2^3 \cdot 3^4 \cdot 5 \cdot 17 \cdot 29 \cdot 41 \cdot 43 \cdot 79 \cdot 83 \cdot 89$$
$$9\,038\,619\,861\,406\,740 = 2^2 \cdot 3^4 \cdot 5 \cdot 11 \cdot 17 \cdot 29 \cdot 41 \cdot 43 \cdot 79 \cdot 83 \cdot 89$$
$$45\,795\,673\,964\,460\,816 = 2^4 \cdot 3^3 \cdot 11 \cdot 17 \cdot 19 \cdot 29 \cdot 41 \cdot 43 \cdot 79 \cdot 83 \cdot 89$$
$$214\,667\,221\,708\,410\,075 = 3^4 \cdot 5^2 \cdot 11 \cdot 17 \cdot 19 \cdot 29 \cdot 41 \cdot 43 \cdot 79 \cdot 83 \cdot 89$$
$$934\,433\,788\,613\,079\,150 = 2 \cdot 3^4 \cdot 5^2 \cdot 11 \cdot 19 \cdot 29 \cdot 37 \cdot 41 \cdot 43 \cdot 79 \cdot 83 \cdot 89$$
$$3\,789\,648\,142\,708\,598\,775 = 3^2 \cdot 5^2 \cdot 11 \cdot 19 \cdot 29 \cdot 37 \cdot 41 \cdot 43 \cdot 73 \cdot 79 \cdot 83 \cdot 89$$
$$14\,360\,771\,909\,211\,532\,200 = 2^3 \cdot 3^4 \cdot 5^2 \cdot 11 \cdot 29 \cdot 37 \cdot 41 \cdot 43 \cdot 73 \cdot 79 \cdot 83 \cdot 89$$
$$50\,980\,740\,277\,700\,939\,310 = 2 \cdot 3^4 \cdot 5 \cdot 11 \cdot 29 \cdot 37 \cdot 41 \cdot 43 \cdot 71 \cdot 73 \cdot 79 \cdot 83 \cdot 89$$
$$169\,935\,800\,925\,669\,797\,700 = 2^2 \cdot 3^3 \cdot 5^2 \cdot 11 \cdot 29 \cdot 37 \cdot 41 \cdot 43 \cdot 71 \cdot 73 \cdot 79 \cdot 83 \cdot 89$$
$$532\,980\,466\,539\,600\,729\,150 = 2 \cdot 3^4 \cdot 5^2 \cdot 23 \cdot 29 \cdot 37 \cdot 41 \cdot 43 \cdot 71 \cdot 73 \cdot 79 \cdot 83 \cdot 89$$
$$1\,575\,768\,335\,856\,210\,851\,400 = 2^3 \cdot 3^4 \cdot 5^2 \cdot 17 \cdot 29 \cdot 37 \cdot 41 \cdot 43 \cdot 71 \cdot 73 \cdot 79 \cdot 83 \cdot 89$$
$$4\,399\,019\,937\,598\,588\,626\,825 = 3^3 \cdot 5^2 \cdot 17 \cdot 29 \cdot 37 \cdot 41 \cdot 43 \cdot 67 \cdot 71 \cdot 73 \cdot 79 \cdot 83 \cdot 89$$
$$11\,613\,412\,635\,260\,273\,974\,818 = 2 \cdot 3^4 \cdot 11 \cdot 17 \cdot 29 \cdot 37 \cdot 41 \cdot 43 \cdot 67 \cdot 71 \cdot 73 \cdot 79 \cdot 83 \cdot 89$$
$$29\,033\,531\,588\,150\,684\,937\,045 = 3^4 \cdot 5 \cdot 11 \cdot 17 \cdot 29 \cdot 37 \cdot 41 \cdot 43 \cdot 67 \cdot 71 \cdot 73 \cdot 79 \cdot 83 \cdot 89$$
$$68\,820\,223\,023\,764\,586\,517\,440 = 2^6 \cdot 3 \cdot 5 \cdot 11 \cdot 17 \cdot 29 \cdot 37 \cdot 41 \cdot 43 \cdot 67 \cdot 71 \cdot 73 \cdot 79 \cdot 83 \cdot 89$$
$$154\,845\,501\,803\,470\,319\,664\,240 = 2^4 \cdot 3^3 \cdot 5 \cdot 11 \cdot 17 \cdot 29 \cdot 37 \cdot 41 \cdot 43 \cdot 67 \cdot 71 \cdot 73 \cdot 79 \cdot 83 \cdot 89$$
$$331\,049\,003\,855\,695\,166\,178\,720 = 2^5 \cdot 3^3 \cdot 5 \cdot 11 \cdot 17 \cdot 31 \cdot 37 \cdot 41 \cdot 43 \cdot 67 \cdot 71 \cdot 73 \cdot 79 \cdot 83 \cdot 89$$
$$673\,132\,974\,506\,580\,171\,230\,064 = 2^4 \cdot 3^2 \cdot 11 \cdot 17 \cdot 31 \cdot 37 \cdot 41 \cdot 43 \cdot 61 \cdot 67 \cdot 71 \cdot 73 \cdot 79 \cdot 83 \cdot 89$$
$$1\,302\,838\,015\,174\,026\,137\,864\,640 = 2^6 \cdot 3^3 \cdot 5 \cdot 11 \cdot 17 \cdot 37 \cdot 41 \cdot 43 \cdot 61 \cdot 67 \cdot 71 \cdot 73 \cdot 79 \cdot 83 \cdot 89$$
$$2\,402\,107\,590\,477\,110\,691\,687\,930 = 2 \cdot 3^3 \cdot 5 \cdot 11 \cdot 17 \cdot 37 \cdot 41 \cdot 43 \cdot 59 \cdot 61 \cdot 67 \cdot 71 \cdot 73 \cdot 79 \cdot 83 \cdot 89$$
$$4\,221\,886\,068\,111\,285\,458\,118\,180 = 2^2 \cdot 3^2 \cdot 5 \cdot 17 \cdot 29 \cdot 37 \cdot 41 \cdot 43 \cdot 59 \cdot 61 \cdot 67 \cdot 71 \cdot 73 \cdot 79 \cdot 83 \cdot 89$$
$$7\,077\,867\,820\,068\,919\,738\,609\,890 = 2 \cdot 3^3 \cdot 5 \cdot 19 \cdot 29 \cdot 37 \cdot 41 \cdot 43 \cdot 59 \cdot 61 \cdot 67 \cdot 71 \cdot 73 \cdot 79 \cdot 83 \cdot 89$$
$$11\,324\,588\,512\,110\,271\,581\,775\,824 = 2^4 \cdot 3^3 \cdot 19 \cdot 29 \cdot 37 \cdot 41 \cdot 43 \cdot 59 \cdot 61 \cdot 67 \cdot 71 \cdot 73 \cdot 79 \cdot 83 \cdot 89$$
$$17\,301\,454\,671\,279\,581\,583\,268\,620 = 2^2 \cdot 3 \cdot 5 \cdot 11 \cdot 19 \cdot 29 \cdot 37 \cdot 41 \cdot 43 \cdot 59 \cdot 61 \cdot 67 \cdot 71 \cdot 73 \cdot 79 \cdot 83 \cdot 89$$
$$25\,250\,771\,682\,408\,037\,986\,392\,040 = 2^3 \cdot 3^4 \cdot 5 \cdot 11 \cdot 19 \cdot 29 \cdot 41 \cdot 43 \cdot 59 \cdot 61 \cdot 67 \cdot 71 \cdot 73 \cdot 79 \cdot 83 \cdot 89$$
$$35\,218\,181\,557\,042\,789\,823\,125\,740 = 2^2 \cdot 3^4 \cdot 5 \cdot 11 \cdot 29 \cdot 41 \cdot 43 \cdot 53 \cdot 59 \cdot 61 \cdot 67 \cdot 71 \cdot 73 \cdot 79 \cdot 83 \cdot 89$$
$$46\,957\,575\,409\,390\,386\,430\,834\,320 = 2^4 \cdot 3^3 \cdot 5 \cdot 11 \cdot 29 \cdot 41 \cdot 43 \cdot 53 \cdot 59 \cdot 61 \cdot 67 \cdot 71 \cdot 73 \cdot 79 \cdot 83 \cdot 89$$
$$59\,870\,908\,646\,972\,742\,699\,313\,758 = 2 \cdot 3^4 \cdot 11 \cdot 17 \cdot 29 \cdot 41 \cdot 43 \cdot 53 \cdot 59 \cdot 61 \cdot 67 \cdot 71 \cdot 73 \cdot 79 \cdot 83 \cdot 89$$
$$73\,013\,303\,228\,015\,539\,877\,211\,900 = 2^2 \cdot 3^4 \cdot 5^2 \cdot 11 \cdot 17 \cdot 29 \cdot 43 \cdot 53 \cdot 59 \cdot 61 \cdot 67 \cdot 71 \cdot 73 \cdot 79 \cdot 83 \cdot 89$$
$$85\,182\,187\,099\,351\,463\,190\,080\,550 = 2 \cdot 3^3 \cdot 5^2 \cdot 7 \cdot 11 \cdot 17 \cdot 29 \cdot 43 \cdot 53 \cdot 59 \cdot 61 \cdot 67 \cdot 71 \cdot 73 \cdot 79 \cdot 83 \cdot 89$$
$$95\,087\,092\,576\,020\,237\,979\,624\,800 = 2^5 \cdot 3^4 \cdot 5^2 \cdot 7 \cdot 11 \cdot 17 \cdot 29 \cdot 53 \cdot 59 \cdot 61 \cdot 67 \cdot 71 \cdot 73 \cdot 79 \cdot 83 \cdot 89$$
$$101\,570\,303\,433\,476\,163\,296\,417\,400 = 2^3 \cdot 3^4 \cdot 5^2 \cdot 7 \cdot 17 \cdot 29 \cdot 47 \cdot 53 \cdot 59 \cdot 61 \cdot 67 \cdot 71 \cdot 73 \cdot 79 \cdot 83 \cdot 89$$
$$103\,827\,421\,287\,553\,411\,369\,671\,120 = 2^4 \cdot 3^2 \cdot 5 \cdot 7 \cdot 17 \cdot 23 \cdot 29 \cdot 47 \cdot 53 \cdot 59 \cdot 61 \cdot 67 \cdot 71 \cdot 73 \cdot 79 \cdot 83 \cdot 89$$

Row 91

$$91 = 7 \cdot 13$$
$$4\,095 = 3^2 \cdot 5 \cdot 7 \cdot 13$$
$$121\,485 = 3 \cdot 5 \cdot 7 \cdot 13 \cdot 89$$
$$2\,672\,670 = 2 \cdot 3 \cdot 5 \cdot 7 \cdot 11 \cdot 13 \cdot 89$$
$$46\,504\,458 = 2 \cdot 3^2 \cdot 7 \cdot 11 \cdot 13 \cdot 29 \cdot 89$$
$$666\,563\,898 = 2 \cdot 3 \cdot 7 \cdot 11 \cdot 13 \cdot 29 \cdot 43 \cdot 89$$
$$8\,093\,990\,190 = 2 \cdot 3 \cdot 5 \cdot 11 \cdot 13 \cdot 17 \cdot 29 \cdot 43 \cdot 89$$
$$84\,986\,896\,995 = 3^2 \cdot 5 \cdot 7 \cdot 11 \cdot 13 \cdot 17 \cdot 29 \cdot 43 \cdot 89$$
$$783\,768\,050\,065 = 5 \cdot 7 \cdot 11 \cdot 13 \cdot 17 \cdot 29 \cdot 43 \cdot 83 \cdot 89$$
$$6\,426\,898\,010\,533 = 7 \cdot 11 \cdot 13 \cdot 17 \cdot 29 \cdot 41 \cdot 43 \cdot 83 \cdot 89$$
$$47\,325\,339\,895\,743 = 3^4 \cdot 7 \cdot 13 \cdot 17 \cdot 29 \cdot 41 \cdot 43 \cdot 83 \cdot 89$$
$$315\,502\,265\,971\,620 = 2^2 \cdot 3^3 \cdot 5 \cdot 7 \cdot 13 \cdot 17 \cdot 29 \cdot 41 \cdot 43 \cdot 83 \cdot 89$$
$$1\,917\,283\,000\,904\,460 = 2^2 \cdot 3^3 \cdot 5 \cdot 7 \cdot 17 \cdot 29 \cdot 41 \cdot 43 \cdot 79 \cdot 83 \cdot 89$$
$$10\,682\,005\,290\,753\,420 = 2^2 \cdot 3^4 \cdot 5 \cdot 13 \cdot 17 \cdot 29 \cdot 41 \cdot 43 \cdot 79 \cdot 83 \cdot 89$$
$$54\,834\,293\,825\,867\,556 = 2^2 \cdot 3^3 \cdot 7 \cdot 11 \cdot 13 \cdot 17 \cdot 29 \cdot 41 \cdot 43 \cdot 79 \cdot 83 \cdot 89$$
$$260\,462\,895\,672\,870\,891 = 3^3 \cdot 7 \cdot 11 \cdot 13 \cdot 17 \cdot 19 \cdot 29 \cdot 41 \cdot 43 \cdot 79 \cdot 83 \cdot 89$$
$$1\,149\,101\,010\,321\,489\,225 = 3^4 \cdot 5^2 \cdot 7 \cdot 11 \cdot 13 \cdot 19 \cdot 29 \cdot 41 \cdot 43 \cdot 79 \cdot 83 \cdot 89$$
$$4\,724\,081\,931\,321\,677\,925 = 3^2 \cdot 5^2 \cdot 7 \cdot 11 \cdot 13 \cdot 19 \cdot 29 \cdot 37 \cdot 41 \cdot 43 \cdot 79 \cdot 83 \cdot 89$$
$$18\,150\,420\,051\,920\,130\,975 = 3^2 \cdot 5^2 \cdot 7 \cdot 11 \cdot 13 \cdot 29 \cdot 37 \cdot 41 \cdot 43 \cdot 73 \cdot 79 \cdot 83 \cdot 89$$
$$65\,341\,512\,186\,912\,471\,510 = 2 \cdot 3^4 \cdot 5 \cdot 7 \cdot 11 \cdot 13 \cdot 29 \cdot 37 \cdot 41 \cdot 43 \cdot 73 \cdot 79 \cdot 83 \cdot 89$$
$$220\,916\,541\,203\,370\,737\,010 = 2 \cdot 3^3 \cdot 5 \cdot 11 \cdot 13 \cdot 29 \cdot 37 \cdot 41 \cdot 43 \cdot 71 \cdot 73 \cdot 79 \cdot 83 \cdot 89$$
$$702\,916\,267\,465\,270\,526\,850 = 2 \cdot 3^3 \cdot 5^2 \cdot 7 \cdot 13 \cdot 29 \cdot 37 \cdot 41 \cdot 43 \cdot 71 \cdot 73 \cdot 79 \cdot 83 \cdot 89$$
$$2\,108\,748\,802\,395\,811\,580\,550 = 2 \cdot 3^4 \cdot 5^2 \cdot 7 \cdot 13 \cdot 29 \cdot 37 \cdot 41 \cdot 43 \cdot 71 \cdot 73 \cdot 79 \cdot 83 \cdot 89$$
$$5\,974\,788\,273\,454\,799\,478\,225 = 3^3 \cdot 5^2 \cdot 7 \cdot 13 \cdot 17 \cdot 29 \cdot 37 \cdot 41 \cdot 43 \cdot 71 \cdot 73 \cdot 79 \cdot 83 \cdot 89$$
$$16\,012\,432\,572\,858\,862\,601\,643 = 3^3 \cdot 7 \cdot 13 \cdot 17 \cdot 29 \cdot 37 \cdot 41 \cdot 43 \cdot 67 \cdot 71 \cdot 73 \cdot 79 \cdot 83 \cdot 89$$
$$40\,646\,944\,223\,410\,958\,911\,863 = 3^4 \cdot 7 \cdot 11 \cdot 17 \cdot 29 \cdot 37 \cdot 41 \cdot 43 \cdot 67 \cdot 71 \cdot 73 \cdot 79 \cdot 83 \cdot 89$$
$$97\,853\,754\,611\,915\,271\,454\,485 = 3 \cdot 5 \cdot 7 \cdot 11 \cdot 13 \cdot 17 \cdot 29 \cdot 37 \cdot 41 \cdot 43 \cdot 67 \cdot 71 \cdot 73 \cdot 79 \cdot 83 \cdot 89$$
$$223\,665\,724\,827\,234\,906\,181\,680 = 2^4 \cdot 3 \cdot 5 \cdot 11 \cdot 13 \cdot 17 \cdot 29 \cdot 37 \cdot 41 \cdot 43 \cdot 67 \cdot 71 \cdot 73 \cdot 79 \cdot 83 \cdot 89$$
$$485\,894\,505\,659\,165\,485\,842\,960 = 2^4 \cdot 3^3 \cdot 5 \cdot 7 \cdot 11 \cdot 13 \cdot 17 \cdot 37 \cdot 41 \cdot 43 \cdot 67 \cdot 71 \cdot 73 \cdot 79 \cdot 83 \cdot 89$$
$$1\,004\,181\,978\,362\,275\,337\,408\,784 = 2^4 \cdot 3^2 \cdot 7 \cdot 11 \cdot 13 \cdot 17 \cdot 31 \cdot 37 \cdot 41 \cdot 43 \cdot 67 \cdot 71 \cdot 73 \cdot 79 \cdot 83 \cdot 89$$
$$1\,975\,970\,989\,680\,606\,309\,094\,704 = 2^4 \cdot 3^2 \cdot 7 \cdot 11 \cdot 13 \cdot 17 \cdot 37 \cdot 41 \cdot 43 \cdot 61 \cdot 67 \cdot 71 \cdot 73 \cdot 79 \cdot 83 \cdot 89$$
$$3\,704\,945\,605\,651\,136\,829\,552\,570 = 2 \cdot 3^3 \cdot 5 \cdot 7 \cdot 11 \cdot 13 \cdot 17 \cdot 37 \cdot 41 \cdot 43 \cdot 61 \cdot 67 \cdot 71 \cdot 73 \cdot 79 \cdot 83 \cdot 89$$
$$6\,623\,993\,658\,588\,396\,149\,806\,110 = 2 \cdot 3^2 \cdot 5 \cdot 7 \cdot 13 \cdot 17 \cdot 37 \cdot 41 \cdot 43 \cdot 59 \cdot 61 \cdot 67 \cdot 71 \cdot 73 \cdot 79 \cdot 83 \cdot 89$$
$$11\,299\,753\,888\,180\,205\,196\,728\,070 = 2 \cdot 3^2 \cdot 5 \cdot 7 \cdot 13 \cdot 29 \cdot 37 \cdot 41 \cdot 43 \cdot 59 \cdot 61 \cdot 67 \cdot 71 \cdot 73 \cdot 79 \cdot 83 \cdot 89$$
$$18\,402\,456\,332\,179\,191\,320\,385\,714 = 2 \cdot 3^3 \cdot 13 \cdot 19 \cdot 29 \cdot 37 \cdot 41 \cdot 43 \cdot 59 \cdot 61 \cdot 67 \cdot 71 \cdot 73 \cdot 79 \cdot 83 \cdot 89$$
$$28\,626\,043\,183\,389\,853\,165\,044\,444 = 2^2 \cdot 3 \cdot 7 \cdot 13 \cdot 19 \cdot 29 \cdot 37 \cdot 41 \cdot 43 \cdot 59 \cdot 61 \cdot 67 \cdot 71 \cdot 73 \cdot 79 \cdot 83 \cdot 89$$
$$42\,552\,226\,353\,687\,619\,569\,660\,660 = 2^2 \cdot 3 \cdot 5 \cdot 7 \cdot 11 \cdot 13 \cdot 19 \cdot 29 \cdot 41 \cdot 43 \cdot 59 \cdot 61 \cdot 67 \cdot 71 \cdot 73 \cdot 79 \cdot 83 \cdot 89$$
$$60\,468\,953\,239\,450\,827\,809\,517\,780 = 2^2 \cdot 3^4 \cdot 5 \cdot 7 \cdot 11 \cdot 13 \cdot 29 \cdot 41 \cdot 43 \cdot 59 \cdot 61 \cdot 67 \cdot 71 \cdot 73 \cdot 79 \cdot 83 \cdot 89$$
$$82\,175\,756\,966\,433\,176\,253\,960\,060 = 2^2 \cdot 3^3 \cdot 5 \cdot 7 \cdot 11 \cdot 29 \cdot 41 \cdot 43 \cdot 53 \cdot 59 \cdot 61 \cdot 67 \cdot 71 \cdot 73 \cdot 79 \cdot 83 \cdot 89$$
$$106\,828\,484\,056\,363\,129\,130\,148\,078 = 2 \cdot 3^3 \cdot 7 \cdot 11 \cdot 13 \cdot 29 \cdot 41 \cdot 43 \cdot 53 \cdot 59 \cdot 61 \cdot 67 \cdot 71 \cdot 73 \cdot 79 \cdot 83 \cdot 89$$
$$132\,884\,211\,874\,988\,282\,576\,525\,658 = 2 \cdot 3^4 \cdot 7 \cdot 11 \cdot 13 \cdot 17 \cdot 29 \cdot 43 \cdot 53 \cdot 59 \cdot 61 \cdot 67 \cdot 71 \cdot 73 \cdot 79 \cdot 83 \cdot 89$$
$$158\,195\,490\,327\,367\,003\,067\,292\,450 = 2 \cdot 3^3 \cdot 5^2 \cdot 11 \cdot 13 \cdot 17 \cdot 29 \cdot 43 \cdot 53 \cdot 59 \cdot 61 \cdot 67 \cdot 71 \cdot 73 \cdot 79 \cdot 83 \cdot 89$$
$$180\,269\,279\,675\,371\,701\,169\,705\,350 = 2 \cdot 3^3 \cdot 5^2 \cdot 7^2 \cdot 11 \cdot 13 \cdot 17 \cdot 29 \cdot 53 \cdot 59 \cdot 61 \cdot 67 \cdot 71 \cdot 73 \cdot 79 \cdot 83 \cdot 89$$
$$196\,657\,396\,009\,496\,401\,276\,042\,200 = 2^3 \cdot 3^4 \cdot 5^2 \cdot 7^2 \cdot 13 \cdot 17 \cdot 29 \cdot 53 \cdot 59 \cdot 61 \cdot 67 \cdot 71 \cdot 73 \cdot 79 \cdot 83 \cdot 89$$
$$205\,397\,724\,721\,029\,574\,666\,088\,520 = 2^3 \cdot 3^2 \cdot 5 \cdot 7^2 \cdot 13 \cdot 17 \cdot 29 \cdot 47 \cdot 53 \cdot 59 \cdot 61 \cdot 67 \cdot 71 \cdot 73 \cdot 79 \cdot 83 \cdot 89$$

Pascal's Triangle – Prime Factorization – To Center Number (omitting 1's)

Row 92

$92 = 2^2 \cdot 23$

$4\,186 = 2 \cdot 7 \cdot 13 \cdot 23$

$125\,580 = 2^2 \cdot 3 \cdot 5 \cdot 7 \cdot 13 \cdot 23$

$2\,794\,155 = 3 \cdot 5 \cdot 7 \cdot 13 \cdot 23 \cdot 89$

$49\,177\,128 = 2^3 \cdot 3 \cdot 7 \cdot 11 \cdot 13 \cdot 23 \cdot 89$

$713\,068\,356 = 2^2 \cdot 3 \cdot 7 \cdot 11 \cdot 13 \cdot 23 \cdot 29 \cdot 89$

$8\,760\,554\,088 = 2^3 \cdot 3 \cdot 11 \cdot 13 \cdot 23 \cdot 29 \cdot 43 \cdot 89$

$93\,080\,887\,185 = 3 \cdot 5 \cdot 11 \cdot 13 \cdot 17 \cdot 23 \cdot 29 \cdot 43 \cdot 89$

$868\,754\,947\,060 = 2^2 \cdot 5 \cdot 7 \cdot 11 \cdot 13 \cdot 17 \cdot 23 \cdot 29 \cdot 43 \cdot 89$

$7\,210\,666\,060\,598 = 2 \cdot 7 \cdot 11 \cdot 13 \cdot 17 \cdot 23 \cdot 29 \cdot 43 \cdot 83 \cdot 89$

$53\,752\,237\,906\,276 = 2^2 \cdot 7 \cdot 13 \cdot 17 \cdot 23 \cdot 29 \cdot 41 \cdot 43 \cdot 83 \cdot 89$

$362\,827\,605\,867\,363 = 3^3 \cdot 7 \cdot 13 \cdot 17 \cdot 23 \cdot 29 \cdot 41 \cdot 43 \cdot 83 \cdot 89$

$2\,232\,785\,266\,876\,080 = 2^4 \cdot 3^3 \cdot 5 \cdot 7 \cdot 17 \cdot 23 \cdot 29 \cdot 41 \cdot 43 \cdot 83 \cdot 89$

$12\,599\,288\,291\,657\,880 = 2^3 \cdot 3^3 \cdot 5 \cdot 17 \cdot 23 \cdot 29 \cdot 41 \cdot 43 \cdot 79 \cdot 83 \cdot 89$

$65\,516\,299\,116\,620\,976 = 2^4 \cdot 3^3 \cdot 13 \cdot 17 \cdot 23 \cdot 29 \cdot 41 \cdot 43 \cdot 79 \cdot 83 \cdot 89$

$315\,297\,189\,498\,738\,447 = 3^3 \cdot 7 \cdot 11 \cdot 13 \cdot 17 \cdot 23 \cdot 29 \cdot 41 \cdot 43 \cdot 79 \cdot 83 \cdot 89$

$1\,409\,563\,905\,994\,360\,116 = 2^2 \cdot 3^3 \cdot 7 \cdot 11 \cdot 13 \cdot 19 \cdot 23 \cdot 29 \cdot 41 \cdot 43 \cdot 79 \cdot 83 \cdot 89$

$5\,873\,182\,941\,643\,167\,150 = 2 \cdot 3^2 \cdot 5^2 \cdot 7 \cdot 11 \cdot 13 \cdot 19 \cdot 23 \cdot 29 \cdot 41 \cdot 43 \cdot 79 \cdot 83 \cdot 89$

$22\,874\,501\,983\,241\,808\,900 = 2^2 \cdot 3^2 \cdot 5^2 \cdot 7 \cdot 11 \cdot 13 \cdot 23 \cdot 29 \cdot 37 \cdot 41 \cdot 43 \cdot 79 \cdot 83 \cdot 89$

$83\,491\,932\,238\,832\,602\,485 = 3^2 \cdot 5 \cdot 7 \cdot 11 \cdot 13 \cdot 23 \cdot 29 \cdot 37 \cdot 41 \cdot 43 \cdot 73 \cdot 79 \cdot 83 \cdot 89$

$286\,258\,053\,390\,283\,208\,520 = 2^3 \cdot 3^3 \cdot 5 \cdot 11 \cdot 13 \cdot 23 \cdot 29 \cdot 37 \cdot 41 \cdot 43 \cdot 73 \cdot 79 \cdot 83 \cdot 89$

$923\,832\,808\,668\,641\,263\,860 = 2^2 \cdot 3^3 \cdot 5 \cdot 13 \cdot 23 \cdot 29 \cdot 37 \cdot 41 \cdot 43 \cdot 71 \cdot 73 \cdot 79 \cdot 83 \cdot 89$

$2\,811\,665\,069\,861\,082\,107\,400 = 2^3 \cdot 3^3 \cdot 5^2 \cdot 7 \cdot 13 \cdot 29 \cdot 37 \cdot 41 \cdot 43 \cdot 71 \cdot 73 \cdot 79 \cdot 83 \cdot 89$

$8\,083\,537\,075\,850\,611\,058\,775 = 3^3 \cdot 5^2 \cdot 7 \cdot 13 \cdot 23 \cdot 29 \cdot 37 \cdot 41 \cdot 43 \cdot 71 \cdot 73 \cdot 79 \cdot 83 \cdot 89$

$21\,987\,220\,846\,313\,662\,079\,868 = 2^2 \cdot 3^3 \cdot 7 \cdot 13 \cdot 17 \cdot 23 \cdot 29 \cdot 37 \cdot 41 \cdot 43 \cdot 71 \cdot 73 \cdot 79 \cdot 83 \cdot 89$

$56\,659\,376\,796\,269\,821\,513\,506 = 2 \cdot 3^3 \cdot 7 \cdot 17 \cdot 23 \cdot 29 \cdot 37 \cdot 41 \cdot 43 \cdot 67 \cdot 71 \cdot 73 \cdot 79 \cdot 83 \cdot 89$

$138\,500\,698\,835\,326\,230\,366\,348 = 2^2 \cdot 3 \cdot 7 \cdot 11 \cdot 17 \cdot 23 \cdot 29 \cdot 37 \cdot 41 \cdot 43 \cdot 67 \cdot 71 \cdot 73 \cdot 79 \cdot 83 \cdot 89$

$321\,519\,479\,439\,150\,177\,636\,165 = 3 \cdot 5 \cdot 11 \cdot 13 \cdot 17 \cdot 23 \cdot 29 \cdot 37 \cdot 41 \cdot 43 \cdot 67 \cdot 71 \cdot 73 \cdot 79 \cdot 83 \cdot 89$

$709\,560\,230\,486\,400\,392\,024\,640 = 2^6 \cdot 3 \cdot 5 \cdot 11 \cdot 13 \cdot 17 \cdot 23 \cdot 37 \cdot 41 \cdot 43 \cdot 67 \cdot 71 \cdot 73 \cdot 79 \cdot 83 \cdot 89$

$1\,490\,076\,484\,021\,440\,823\,251\,744 = 2^5 \cdot 3^2 \cdot 7 \cdot 11 \cdot 13 \cdot 17 \cdot 23 \cdot 37 \cdot 41 \cdot 43 \cdot 67 \cdot 71 \cdot 73 \cdot 79 \cdot 83 \cdot 89$

$2\,980\,152\,968\,042\,881\,646\,503\,488 = 2^6 \cdot 3^2 \cdot 7 \cdot 11 \cdot 13 \cdot 17 \cdot 23 \cdot 37 \cdot 41 \cdot 43 \cdot 67 \cdot 71 \cdot 73 \cdot 79 \cdot 83 \cdot 89$

$5\,680\,916\,595\,331\,743\,138\,647\,274 = 2 \cdot 3^2 \cdot 7 \cdot 11 \cdot 13 \cdot 17 \cdot 23 \cdot 37 \cdot 41 \cdot 43 \cdot 61 \cdot 67 \cdot 71 \cdot 73 \cdot 79 \cdot 83 \cdot 89$

$10\,328\,939\,264\,239\,532\,979\,358\,680 = 2^3 \cdot 3^2 \cdot 5 \cdot 7 \cdot 13 \cdot 17 \cdot 23 \cdot 37 \cdot 41 \cdot 43 \cdot 61 \cdot 67 \cdot 71 \cdot 73 \cdot 79 \cdot 83 \cdot 89$

$17\,923\,747\,546\,768\,601\,346\,534\,180 = 2^2 \cdot 3^2 \cdot 5 \cdot 7 \cdot 13 \cdot 23 \cdot 37 \cdot 41 \cdot 43 \cdot 59 \cdot 61 \cdot 67 \cdot 71 \cdot 73 \cdot 79 \cdot 83 \cdot 89$

$29\,702\,210\,220\,359\,396\,517\,113\,784 = 2^3 \cdot 3^2 \cdot 13 \cdot 23 \cdot 29 \cdot 37 \cdot 41 \cdot 43 \cdot 59 \cdot 61 \cdot 67 \cdot 71 \cdot 73 \cdot 79 \cdot 83 \cdot 89$

$47\,028\,499\,515\,569\,044\,485\,430\,158 = 2 \cdot 3 \cdot 13 \cdot 19 \cdot 23 \cdot 29 \cdot 37 \cdot 41 \cdot 43 \cdot 59 \cdot 61 \cdot 67 \cdot 71 \cdot 73 \cdot 79 \cdot 83 \cdot 89$

$71\,178\,269\,537\,077\,472\,734\,705\,104 = 2^4 \cdot 3 \cdot 7 \cdot 13 \cdot 19 \cdot 23 \cdot 29 \cdot 41 \cdot 43 \cdot 59 \cdot 61 \cdot 67 \cdot 71 \cdot 73 \cdot 79 \cdot 83 \cdot 89$

$103\,021\,179\,593\,138\,447\,379\,178\,440 = 2^3 \cdot 3 \cdot 5 \cdot 7 \cdot 11 \cdot 13 \cdot 23 \cdot 29 \cdot 41 \cdot 43 \cdot 59 \cdot 61 \cdot 67 \cdot 71 \cdot 73 \cdot 79 \cdot 83 \cdot 89$

$142\,644\,710\,205\,884\,004\,063\,477\,840 = 2^4 \cdot 3^3 \cdot 5 \cdot 7 \cdot 11 \cdot 23 \cdot 29 \cdot 41 \cdot 43 \cdot 59 \cdot 61 \cdot 67 \cdot 71 \cdot 73 \cdot 79 \cdot 83 \cdot 89$

$189\,004\,241\,022\,796\,305\,384\,108\,138 = 2 \cdot 3^3 \cdot 7 \cdot 11 \cdot 23 \cdot 29 \cdot 41 \cdot 43 \cdot 53 \cdot 59 \cdot 61 \cdot 67 \cdot 71 \cdot 73 \cdot 79 \cdot 83 \cdot 89$

$239\,712\,695\,931\,351\,411\,706\,673\,736 = 2^3 \cdot 3^3 \cdot 7 \cdot 11 \cdot 13 \cdot 23 \cdot 29 \cdot 43 \cdot 53 \cdot 59 \cdot 61 \cdot 67 \cdot 71 \cdot 73 \cdot 79 \cdot 83 \cdot 89$

$291\,079\,702\,202\,355\,285\,643\,818\,108 = 2^2 \cdot 3^3 \cdot 11 \cdot 13 \cdot 17 \cdot 23 \cdot 29 \cdot 43 \cdot 53 \cdot 59 \cdot 61 \cdot 67 \cdot 71 \cdot 73 \cdot 79 \cdot 83 \cdot 89$

$338\,464\,770\,002\,738\,704\,236\,997\,800 = 2^3 \cdot 3^3 \cdot 5^2 \cdot 11 \cdot 13 \cdot 17 \cdot 23 \cdot 29 \cdot 53 \cdot 59 \cdot 61 \cdot 67 \cdot 71 \cdot 73 \cdot 79 \cdot 83 \cdot 89$

$376\,926\,675\,684\,868\,102\,445\,747\,550 = 2 \cdot 3^3 \cdot 5^2 \cdot 7^2 \cdot 13 \cdot 17 \cdot 23 \cdot 29 \cdot 53 \cdot 59 \cdot 61 \cdot 67 \cdot 71 \cdot 73 \cdot 79 \cdot 83 \cdot 89$

$402\,055\,120\,730\,525\,975\,942\,130\,720 = 2^5 \cdot 3^2 \cdot 5 \cdot 7^2 \cdot 13 \cdot 17 \cdot 23 \cdot 29 \cdot 53 \cdot 59 \cdot 61 \cdot 67 \cdot 71 \cdot 73 \cdot 79 \cdot 83 \cdot 89$

$410\,795\,449\,442\,059\,149\,332\,177\,040 = 2^4 \cdot 3^2 \cdot 5 \cdot 7^2 \cdot 13 \cdot 17 \cdot 29 \cdot 47 \cdot 53 \cdot 59 \cdot 61 \cdot 67 \cdot 71 \cdot 73 \cdot 79 \cdot 83 \cdot 89$

Pascal's Triangle – Prime Factorization – To Center Number (omitting 1's)

Row 93

$$93 = 3 \cdot 31$$

$$4\,278 = 2 \cdot 3 \cdot 23 \cdot 31$$

$$129\,766 = 2 \cdot 7 \cdot 13 \cdot 23 \cdot 31$$

$$2\,919\,735 = 3^2 \cdot 5 \cdot 7 \cdot 13 \cdot 23 \cdot 31$$

$$51\,971\,283 = 3^2 \cdot 7 \cdot 13 \cdot 23 \cdot 31 \cdot 89$$

$$762\,245\,484 = 2^2 \cdot 3 \cdot 7 \cdot 11 \cdot 13 \cdot 23 \cdot 31 \cdot 89$$

$$9\,473\,622\,444 = 2^2 \cdot 3^2 \cdot 11 \cdot 13 \cdot 23 \cdot 29 \cdot 31 \cdot 89$$

$$101\,841\,441\,273 = 3^2 \cdot 11 \cdot 13 \cdot 23 \cdot 29 \cdot 31 \cdot 43 \cdot 89$$

$$961\,835\,834\,245 = 5 \cdot 11 \cdot 13 \cdot 17 \cdot 23 \cdot 29 \cdot 31 \cdot 43 \cdot 89$$

$$8\,079\,421\,007\,658 = 2 \cdot 3 \cdot 7 \cdot 11 \cdot 13 \cdot 17 \cdot 23 \cdot 29 \cdot 31 \cdot 43 \cdot 89$$

$$60\,962\,903\,966\,874 = 2 \cdot 3 \cdot 7 \cdot 13 \cdot 17 \cdot 23 \cdot 29 \cdot 31 \cdot 43 \cdot 83 \cdot 89$$

$$416\,579\,843\,773\,639 = 7 \cdot 13 \cdot 17 \cdot 23 \cdot 29 \cdot 31 \cdot 41 \cdot 43 \cdot 83 \cdot 89$$

$$2\,595\,612\,872\,743\,443 = 3^4 \cdot 7 \cdot 17 \cdot 23 \cdot 29 \cdot 31 \cdot 41 \cdot 43 \cdot 83 \cdot 89$$

$$14\,832\,073\,558\,533\,960 = 2^3 \cdot 3^4 \cdot 5 \cdot 17 \cdot 23 \cdot 29 \cdot 31 \cdot 41 \cdot 43 \cdot 83 \cdot 89$$

$$78\,115\,587\,408\,278\,856 = 2^3 \cdot 3^3 \cdot 17 \cdot 23 \cdot 29 \cdot 31 \cdot 41 \cdot 43 \cdot 79 \cdot 83 \cdot 89$$

$$380\,813\,488\,615\,359\,423 = 3^4 \cdot 13 \cdot 17 \cdot 23 \cdot 29 \cdot 31 \cdot 41 \cdot 43 \cdot 79 \cdot 83 \cdot 89$$

$$1\,724\,861\,095\,493\,098\,563 = 3^4 \cdot 7 \cdot 11 \cdot 13 \cdot 23 \cdot 29 \cdot 31 \cdot 41 \cdot 43 \cdot 79 \cdot 83 \cdot 89$$

$$7\,282\,746\,847\,637\,527\,266 = 2 \cdot 3^2 \cdot 7 \cdot 11 \cdot 13 \cdot 19 \cdot 23 \cdot 29 \cdot 31 \cdot 41 \cdot 43 \cdot 79 \cdot 83 \cdot 89$$

$$28\,747\,684\,924\,884\,976\,050 = 2 \cdot 3^3 \cdot 5^2 \cdot 7 \cdot 11 \cdot 13 \cdot 23 \cdot 29 \cdot 31 \cdot 41 \cdot 43 \cdot 79 \cdot 83 \cdot 89$$

$$106\,366\,434\,222\,074\,411\,385 = 3^3 \cdot 5 \cdot 7 \cdot 11 \cdot 13 \cdot 23 \cdot 29 \cdot 31 \cdot 37 \cdot 41 \cdot 43 \cdot 79 \cdot 83 \cdot 89$$

$$369\,749\,985\,629\,115\,811\,005 = 3^2 \cdot 5 \cdot 11 \cdot 13 \cdot 23 \cdot 29 \cdot 31 \cdot 37 \cdot 41 \cdot 43 \cdot 73 \cdot 79 \cdot 83 \cdot 89$$

$$1\,210\,090\,862\,058\,924\,472\,380 = 2^2 \cdot 3^4 \cdot 5 \cdot 13 \cdot 23 \cdot 29 \cdot 31 \cdot 37 \cdot 41 \cdot 43 \cdot 73 \cdot 79 \cdot 83 \cdot 89$$

$$3\,735\,497\,878\,529\,723\,371\,260 = 2^2 \cdot 3^4 \cdot 5 \cdot 13 \cdot 29 \cdot 31 \cdot 37 \cdot 41 \cdot 43 \cdot 71 \cdot 73 \cdot 79 \cdot 83 \cdot 89$$

$$10\,895\,202\,145\,711\,693\,166\,175 = 3^3 \cdot 5^2 \cdot 7 \cdot 13 \cdot 29 \cdot 31 \cdot 37 \cdot 41 \cdot 43 \cdot 71 \cdot 73 \cdot 79 \cdot 83 \cdot 89$$

$$30\,070\,757\,922\,164\,273\,138\,643 = 3^4 \cdot 7 \cdot 13 \cdot 23 \cdot 29 \cdot 31 \cdot 37 \cdot 41 \cdot 43 \cdot 71 \cdot 73 \cdot 79 \cdot 83 \cdot 89$$

$$78\,646\,597\,642\,583\,483\,593\,374 = 2 \cdot 3^4 \cdot 7 \cdot 17 \cdot 23 \cdot 29 \cdot 31 \cdot 37 \cdot 41 \cdot 43 \cdot 71 \cdot 73 \cdot 79 \cdot 83 \cdot 89$$

$$195\,160\,075\,631\,596\,051\,879\,854 = 2 \cdot 3 \cdot 7 \cdot 17 \cdot 23 \cdot 29 \cdot 31 \cdot 37 \cdot 41 \cdot 43 \cdot 67 \cdot 71 \cdot 73 \cdot 79 \cdot 83 \cdot 89$$

$$460\,020\,178\,274\,476\,408\,002\,513 = 3^2 \cdot 11 \cdot 17 \cdot 23 \cdot 29 \cdot 31 \cdot 37 \cdot 41 \cdot 43 \cdot 67 \cdot 71 \cdot 73 \cdot 79 \cdot 83 \cdot 89$$

$$1\,031\,079\,709\,925\,550\,569\,660\,805 = 3^2 \cdot 5 \cdot 11 \cdot 13 \cdot 17 \cdot 23 \cdot 31 \cdot 37 \cdot 41 \cdot 43 \cdot 67 \cdot 71 \cdot 73 \cdot 79 \cdot 83 \cdot 89$$

$$2\,199\,636\,714\,507\,841\,215\,276\,384 = 2^5 \cdot 3 \cdot 11 \cdot 13 \cdot 17 \cdot 23 \cdot 31 \cdot 37 \cdot 41 \cdot 43 \cdot 67 \cdot 71 \cdot 73 \cdot 79 \cdot 83 \cdot 89$$

$$4\,470\,229\,452\,064\,322\,469\,755\,232 = 2^5 \cdot 3^3 \cdot 7 \cdot 11 \cdot 13 \cdot 17 \cdot 23 \cdot 37 \cdot 41 \cdot 43 \cdot 67 \cdot 71 \cdot 73 \cdot 79 \cdot 83 \cdot 89$$

$$8\,661\,069\,563\,374\,624\,785\,150\,762 = 2 \cdot 3^3 \cdot 7 \cdot 11 \cdot 13 \cdot 17 \cdot 23 \cdot 31 \cdot 37 \cdot 41 \cdot 43 \cdot 67 \cdot 71 \cdot 73 \cdot 79 \cdot 83 \cdot 89$$

$$16\,009\,855\,859\,571\,276\,118\,005\,954 = 2 \cdot 3^2 \cdot 7 \cdot 13 \cdot 17 \cdot 23 \cdot 31 \cdot 37 \cdot 41 \cdot 43 \cdot 61 \cdot 67 \cdot 71 \cdot 73 \cdot 79 \cdot 83 \cdot 89$$

$$28\,252\,686\,811\,008\,134\,325\,892\,860 = 2^2 \cdot 3^3 \cdot 5 \cdot 7 \cdot 13 \cdot 23 \cdot 31 \cdot 37 \cdot 41 \cdot 43 \cdot 61 \cdot 67 \cdot 71 \cdot 73 \cdot 79 \cdot 83 \cdot 89$$

$$47\,625\,957\,767\,127\,997\,863\,647\,964 = 2^2 \cdot 3^3 \cdot 13 \cdot 23 \cdot 31 \cdot 37 \cdot 41 \cdot 43 \cdot 59 \cdot 61 \cdot 67 \cdot 71 \cdot 73 \cdot 79 \cdot 83 \cdot 89$$

$$76\,730\,709\,735\,928\,441\,002\,543\,942 = 2 \cdot 3 \cdot 13 \cdot 23 \cdot 29 \cdot 31 \cdot 37 \cdot 41 \cdot 43 \cdot 59 \cdot 61 \cdot 67 \cdot 71 \cdot 73 \cdot 79 \cdot 83 \cdot 89$$

$$118\,206\,769\,052\,646\,517\,220\,135\,262 = 2 \cdot 3^2 \cdot 13 \cdot 19 \cdot 23 \cdot 29 \cdot 31 \cdot 41 \cdot 43 \cdot 59 \cdot 61 \cdot 67 \cdot 71 \cdot 73 \cdot 79 \cdot 83 \cdot 89$$

$$174\,199\,449\,130\,215\,920\,113\,883\,544 = 2^3 \cdot 3^2 \cdot 7 \cdot 13 \cdot 23 \cdot 29 \cdot 31 \cdot 41 \cdot 43 \cdot 59 \cdot 61 \cdot 67 \cdot 71 \cdot 73 \cdot 79 \cdot 83 \cdot 89$$

$$245\,665\,889\,799\,022\,451\,442\,656\,280 = 2^3 \cdot 3 \cdot 5 \cdot 7 \cdot 11 \cdot 23 \cdot 29 \cdot 31 \cdot 41 \cdot 43 \cdot 59 \cdot 61 \cdot 67 \cdot 71 \cdot 73 \cdot 79 \cdot 83 \cdot 89$$

$$331\,648\,951\,228\,680\,309\,447\,585\,978 = 2 \cdot 3^4 \cdot 7 \cdot 11 \cdot 23 \cdot 29 \cdot 31 \cdot 41 \cdot 43 \cdot 59 \cdot 61 \cdot 67 \cdot 71 \cdot 73 \cdot 79 \cdot 83 \cdot 89$$

$$428\,716\,936\,954\,147\,717\,090\,781\,874 = 2 \cdot 3^4 \cdot 7 \cdot 11 \cdot 23 \cdot 29 \cdot 31 \cdot 43 \cdot 53 \cdot 59 \cdot 61 \cdot 67 \cdot 71 \cdot 73 \cdot 79 \cdot 83 \cdot 89$$

$$530\,792\,398\,133\,706\,697\,350\,491\,844 = 2^2 \cdot 3^3 \cdot 11 \cdot 13 \cdot 23 \cdot 29 \cdot 31 \cdot 43 \cdot 53 \cdot 59 \cdot 61 \cdot 67 \cdot 71 \cdot 73 \cdot 79 \cdot 83 \cdot 89$$

$$629\,544\,472\,205\,093\,989\,880\,815\,908 = 2^2 \cdot 3^4 \cdot 11 \cdot 13 \cdot 17 \cdot 23 \cdot 29 \cdot 31 \cdot 53 \cdot 59 \cdot 61 \cdot 67 \cdot 71 \cdot 73 \cdot 79 \cdot 83 \cdot 89$$

$$715\,391\,445\,687\,606\,806\,682\,745\,350 = 2 \cdot 3^4 \cdot 5^2 \cdot 13 \cdot 17 \cdot 23 \cdot 29 \cdot 31 \cdot 53 \cdot 59 \cdot 61 \cdot 67 \cdot 71 \cdot 73 \cdot 79 \cdot 83 \cdot 89$$

$$778\,981\,796\,415\,394\,078\,387\,878\,270 = 2 \cdot 3^2 \cdot 5 \cdot 7^2 \cdot 13 \cdot 17 \cdot 23 \cdot 29 \cdot 31 \cdot 53 \cdot 59 \cdot 61 \cdot 67 \cdot 71 \cdot 73 \cdot 79 \cdot 83 \cdot 89$$

$$812\,850\,570\,172\,585\,125\,274\,307\,760 = 2^4 \cdot 3^3 \cdot 5 \cdot 7^2 \cdot 13 \cdot 17 \cdot 29 \cdot 31 \cdot 53 \cdot 59 \cdot 61 \cdot 67 \cdot 71 \cdot 73 \cdot 79 \cdot 83 \cdot 89$$